XIII International Conference on Hole Burning, Single Molecule, and Related Spectroscopies (HBSM-2018)

Science and Applications

EPJ Web of Conferences Volume 190 (2018)

Moscow, Russia
6 – 12 August 2018

Editors:

A.V. Naumov
M.G. Gladush
K.R. Karimullin

ISBN: 978-1-5108-7253-0

TABLE OF CONTENTS

EPJ Web of Conferences **190**, 00001 (2018) https://doi.org/10.1051/epjconf/201819000001
HBSM-2018

XIII International Conference on Hole Burning, Single Molecule, and Related Spectroscopies: Science and Applications (HBSM-2018): preface

Andrei Naumov[*], *Maxim* Gladush, and *Kamil* Karimullin

[1]Institute for Spectroscopy RAS, 108840 Moscow, Troitsk, Russia
[2]Moscow State Pedagogical University, 119991 Moscow, Russia

XIII International Conference on
HOLE BURNING, SINGLE MOLECULE AND RELATED SPECTROSCOPIES: SCIENCE AND APPLICATIONS
August 6 - 12, 2018 Suzdal - Moscow, Russia

HBSM-2018

Foreword

XIII International Conference on Hole Burning, Single Molecule, and Related Spectroscopies: Science and Applications (HBSM-2018) was organized by the Institute for Spectroscopy of the Russian Academy of Sciences (ISAN; Troitsk, Moscow) with the Moscow State Pedagogical University (MSPU) in Suzdal from 6th till 10th August 2018, and in Moscow and Troitsk on 11th August 2018.

Laser selective spectroscopy has a particular significance and value for the Russian scientific community. The emergence of quasi-linear spectra of complex organic compounds in specially selected solvents at low cryogenic temperatures known as "Shpol'skii matrices" was first observed in 1952 by the members of Prof. Eduard Shpol'skii's group from the Department of Theoretical Physics of the Moscow State Pedagogical Institute. In 1972, a former student of Eduard Shpol'skii, Prof. Roman Personov, and his colleagues at the Institute for Spectroscopy of the USSR Academy of Sciences demonstrated successful site-selective laser excitation of narrow luminescence spectra in organic molecules. Two years later, in 1974, Personov's group with their colleagues from the Institute of Physics in Tartu, Estonia (USSR) conducted the fundamental experiments on the persistent spectral hole burning. That work formed the basis of the laser site-selective spectroscopy of molecules which later evolved in single-molecule spectroscopy.

HBSM was known under a few different names. It started in 1987 in Tallin, Estonia (USSR) as "All-Union Symposium on Modern Methods of Laser Spectroscopy of Molecules in Low-Temperature Media". The second meeting took place in Monterey (California, USA, 1991) as "Persistent Spectral Hole Burning: Science and Applications" and the third meeting became "Spectral Hole Burning and Luminescence Line Narrowing:

[*] Corresponding author: naumov@isan.troitsk.ru; web-page: www.single-molecule.ru

Science and Applications" and moved to Ascona (Switzerland, 1992). Later it was organized as "International conferences on Hole-Burning and Related Spectroscopies (HBRS): Science and Applications" in Tokyo (Japan, 1994), Brainerd (Minnesota, USA, 1996) and Hourtin (France, 1999). In 2001, the conference finally found its current name "Hole Burning, Single Molecule, and Related Spectroscopies (HBSM): Science and Applications". It paved its path through Taipei (Taiwan, 2001), Bozeman (Montana, USA, 2003), Aussois (France, 2006), Palm Cove (Australia, 2009), Tübingen (Germany, 2012) and Tartu (Estonia, 2015). This conference series has been exceptionally successful in bringing together participants from the leading research teams around the world to discuss the latest developments in the fundamental science and the applications of site-selective and single-molecule spectroscopies and numerous related topics (condensed matter physics and physical chemistry, nanoscopy, quantum dots, rear-earth ions, nanodiamonds, biophysics, medical nanodiagnostics, quantum technologies, photovoltaics, on-chip technologies).

The scientific program of the conference included 3 plenary lectures, 19 invited talks, 34 oral talks, 41 poster talks, authored in total by 189 scientists from 21 countries (Australia, Belarus, Belgium, Canada, China, Denmark, Estonia, France, Germany, India, Israel, Italy, Japan, Netherlands, Poland, Russia, Serbia, Sweden, Switzerland, Ukraine, USA).

The students and young scientists at HBSM-2018 displayed their research findings in two poster sections while some were selected to give oral presentations. Their participation included a competition for the Best Student/Young Scientist Talk Award. The winners of the awards were Andrey Anisimov (Ioffe Institute RAS, St. Petersburg, Russia) for poster talk and Alexey Tiranov (University of Geneva, Switzerland) for oral talk. As an additional side-event for the young scientists there was a meeting organized by the "Moscow Pedagogical State University Young Minds Section" of the European Physical Society.

The HBSM-2018 was accompanied with the exhibition organized by the sponsoring companies, where the modern equipment for optics, spectroscopy and related areas from the world-wide leading manufacturers was demonstrated. The organizers acknowledge financial support from the Russian Foundation for Basic Research (project No 18-02-20081; www.rfbr.ru) and hi-tech companies working in the field of development, manufacturing and marketing of modern scientific equipment: Educational Lasers – eLas (laser and photonics teaching systems; Freiburg, Germany; www.e-las.com); Technoinfo Ltd (analytical, laboratory and technological equipment; Moscow, Russia; www.technoinfo.co.uk) which is official representative of PicoQuant (pulsed diode lasers, time-resolved data acquisition, single photon counting, and fluorescence instrumentation; Berlin, Germany; www.picoquant.com) in Russia; Superconducting nanotechnology (single photon detector systems for VIS and NIR spectral range; Moscow, Russia; www.scontel.ru); NT-MDT Spectrum Instruments (atomic force microscopes and its combinations with ultrahigh resolution spectroscopy for nanotechnology and applications; Moscow, Russia; www.ntmdt-si.com); Avesta Ltd (lasers and optical components; Moscow, Russia; www.avesta.ru).

This issue of the EPJ Web of Conferences has collected the brief information about HBSM-2018 as well as the summaries of plenary lectures, conference talks and poster presentations. The issue is recommended for specialists who work in the field optics and spectroscopy, as well as for M.S. and Ph.D. students. The full program and details about the conference can be found on the web-site www.hbsm2018.ru). The next XIV International Conference on Hole Burning, Single Molecule, and Related Spectroscopies: Science and Applications (HBSM-2021) will be in Banz Abbey (Bad Staffelstein, Bavaria) in Germany.

International Advisory and Program Committee

Thomas Basché	Johannes Gutenberg University Mainz, Germany
Christian von Borczyskowski	Chemnitz University of Technology, Germany
Marc Brecht	Eberhard Karls Universität Tübingen, Germany

Thierry Chanelière	Paris-Sud University, France
Jean-Pierre Galaup	Paris-Sud University, France
Ryszard Jankowiak	Kansas State University, Manhattan, USA
Fedor Jelezko	Ulm University, Germany
Jaak Kikas	University of Tartu, Estonia
Jürgen Köhler	University of Bayreuth, Germany
H. Peter Lu	Bowling Green State University, USA
Sebastian Mackowski	Nicolaus Copernicus University, Toruń, Poland
Neil Manson	Australian National University, Canberra, Australia
Alfred J. Meixner	Eberhard Karls Universität Tübingen, Germany
William E. Moerner	Stanford University, USA
Norio Murase	Natl Inst. of Adv. Industrial Sci. and Techn., Osaka, Japan
Andrei V. Naumov	Institute for Spectroscopy RAS / MSPU, Moscow, Russia
Michel Orrit	Leiden University, Netherlands
Taras Plakhotnik	Queensland University, Brisbane, Australia
Aleksander Rebane	Montana State University, Bozeman, USA
Vahid Sandoghdar	Max Planck Inst. for the Sci. of Light, Erlangen, Germany
Ivan Scheblykin	Lund University, Sweden
Vinod Subramaniam	University of Twente, Enschede, Netherlands
Toshiro Tani	Tokyo Uni. of Agriculture and Technology, Tsukuba, Japan
Charles Thiel	Montana State University, Bozeman, USA
Niek van Hulst	The Institute of Photonic Sciences, Barcelona, Spain
Martin Vacha	Tokyo Institute of Technology, Japan
Silvia Völker	Leiden University, Netherlands

Local Organizing Committee

HBSM-2018 was managed by the Interinstitutional scientific team working in laser selective spectroscopy and nanoscopy of single molecules, condensed matter and nanostructures (www.single-molecule.ru).

Andrei Naumov (head)	Institute for Spectroscopy RAS / MSPU / MIPT
Kamil Karimullin (sci. secretary)	Institute for Spectroscopy RAS / MSPU
Victor Zadkov (ISAN section chairman)	Institute for Spectroscopy RAS / HSE / MSU
Dmitry Isaev (MSPU section chairman)	Moscow State Pedagogical University
Artem Arzhanov	Institute for Spectroscopy RAS / MSPU
Ivan Eremchev	Institute for Spectroscopy RAS / MIPT
Anna Es'kova	Moscow State Pedagogical University
Svetlana Gladenkova	Moscow State Pedagogical University
Maxim Gladush	Institute for Spectroscopy RAS / MSPU
Alina Golovanova	Institute for Spectroscopy RAS / MSPU
Alexey Gorshelev	Institute for Spectroscopy RAS
Natalia Lozing	ISAN / Higher School of Economics
Konstantin Magaryan	Moscow State Pedagogical University / ISAN
Olga Morozova	Moscow State Pedagogical University
Alexander Ruzaev	Moscow State Pedagogical University
Alexander Savostianov	Institute for Spectroscopy RAS

Volunteers: Irina Alimkina (MSPU); Mikhail Gubin (VSU, MSPU); Maria Knyazeva (MSU); Elizaveta Kozhina (MSPU); Artur Nelubov (MIPT); Sergey Orlov (ISAN); Aleksei Prokhorov (VSU); Anna Pryakhina (MSPU); Eduard Podshivailov (MSU); Alyona Samsonova (MSPU); Ekaterina Smirnova (MIPT); Aleksandra Triznova (MSPU).

The Shpol'skii-Rebane-Personov Prize for outstanding scientific achievements in the area of site-selective spectroscopy

The International Advisory and Program Committee of the International Conference on Hole Burning and Single Molecule Spectroscopy has established a special Prize and a medal named after Prof. Eduard Shpol'skii, Prof. Karl Rebane, and Prof. Roman Personov that will be awarded for outstanding scientific achievements in the area of site-selective spectroscopy. The Prize bears the names of three outstanding scientists – the founders of the scientific direction of selective spectroscopy of complex organic molecules.

Prof. Eduard Shpol'skii and his colleagues from MPSU in 1952 discovered the effect of quasilinear spectra of organic chromophore molecules in the substances later named "Shpol'skii matrices" at cryogenic temperatures, which opened the way for the fine-structure spectral analysis of a huge number of complex organic compounds. The term "Shpol'skii effect" has now become a commonly used term in scientific communities around the world.

Prof. Karl Rebane, who headed the Institute of Physics of the Academy of Sciences of the Estonian SSR, and his colleagues made a great contribution to understanding the appearance the zero phonon spectral lines of single impurity molecules. He developed the theoretical foundations of optical spectroscopy of impurity centers, which subsequently led to the discovery of the narrow stable hole burning simultaneously in the Institute of Physics in Tartu and the Institute for Spectroscopy, USSR Academy of Sciences in 1974. Karl Rebane predicted and then discovered hot luminescence in crystals.

Prof. Roman Personov and his colleagues at the Institute for Spectroscopy of the USSR Academy of Sciences in 1972 were among the first to conduct unique experiments on laser narrowing of fluorescence lines in impurity organic compounds. Many authors both in Russia and worldwide call the effect of laser excitation of the fine-structure luminescence spectra the "Personov effect".

The discoveries of these three scientists stimulated numerous studies in the area of site-selective spectroscopy all over the world, culminating in the detection of spectra of single molecules, which later played a decisive role in the emergence of a new interdisciplinary direction of ultra-high spatial resolution optical spectroscopy, now known as nanoscopy. Numerous investigations in this field have been the basis and the reasons for organizing the HBSM conference series.

Shpol'skii–Rebane–Personov Prize is awarded for the first time at HBSM-2018 to **Prof. Fedor Jelezko** (Ulm University, Germany) *for his pioneering experiments on quantum optics with single molecules, on NV color centers in diamonds and their applications for magnetometry and quantum information*. Fedor Jelezko made full use of zero-phonon line of organic molecules and of color centers to test quantum optical concepts such as light shift, Rabi resonances, and to develop new applications for sensitive magnetometry. Prof. F. Jelezko has made ground-breaking contributions that have established atomic scale lattice defects (color centers) in diamond as a technology that can harness quantum coherence and entanglement even under ambient conditions. Furthermore, he has developed applications in such devices that are ranging from quantum information processing to the realization of new quantum sensing and imaging technologies.

HBSM-2018 Local Organizing Committee

Photo from the EPS Young Minds meeting on HBSM-2018

Photo of the HBSM-2018 participants. August 6, 2018

edp sciences

Web of Conferences

Statement of Peer review

In submitting conference proceedings to *Web of Conferences*, the editors of the proceedings certify to the Publisher that

1. They adhere to its **Policy on Publishing Integrity** in order to safeguard good scientific practice in publishing.
2. All articles have been subjected to peer review administered by the proceedings editors.
3. Reviews have been conducted by expert referees, who have been requested to provide unbiased and constructive comments aimed, whenever possible, at improving the work.
4. Proceedings editors have taken all reasonable steps to ensure the quality of the materials they publish and their decision to accept or reject a paper for publication has been based only on the merits of the work and the relevance to the journal.

Title, date and place of the conference

XIII International Conference on Hole Burning, Single Molecule, and Related Spectroscopies: Science and Applications (HBSM - 2018)
August 6 – 12, 2018; Suzdal – Moscow, Russia

Proceedings editor(s):

Prof. Dr. Andrei V. Naumov
Dr. Maxim G. Gladush
Dr. Kamil R. Karimullin

Date and editor's signature

September 10, 2018

EPJ Web of Conferences **190**, 01001 (2018) https://doi.org/10.1051/epjconf/201819001001
HBSM-2018

The story of single molecules, from hole-burning and early FM spectroscopy in solids, to super-resolution nanoscopy in cells and beyond

William E. Moerner*

Departments of Chemistry, and by courtesy, of Applied Physics, Stanford University, Stanford, CA USA 94305

In the late 1970s and 1980s, many researchers around the world were exploring a novel optical storage concept, frequency domain optical storage, which was based on spectral hole-burning in the zero-phonon lines of impurities in solids at low temperatures. In 1989, experiments aimed at establishing the ultimate limits this optical storage idea led to the first optical detection and spectroscopy of a single molecule in the condensed phase using laser FM spectroscopy. At the unexplored ultimate limit of individual molecules, many surprises occurred where single molecules showed both spontaneous changes (blinking due to spectral diffusion) and light-driven control of emission, properties that were subsequently also observed in 1997 at room temperature with single green fluorescent protein variants. In 2006, PALM and related approaches showed that Abbe's optical diffraction limit of ~200 nm can be circumvented to achieve super-resolution fluorescence microscopy, or nanoscopy, with relatively nonperturbative visible light. Essential to this advance is the combination of single-molecule fluorescence imaging with active control of the emitting concentration and sequential localization of single fluorophores decorating a structure. Super-resolution microscopy has opened up a new frontier in which biological structures and behavior can be observed in live cells with resolutions down to 20-40 nm and below. Examples range from protein superstructures in bacteria to bands in axons to details of the shapes of amyloid fibrils and much more. Current methods development research addresses ways to extract more information from each single molecule such as 3D position and orientation, and to assure not only precision, but also accuracy. Still, it is worth noting that in spite of all the interest in super-resolution microscopy of extended structures, even in the "conventional" single-molecule tracking regime where the motions of individual biomolecules are recorded in solution or in materials, much can be learned about nanoscale processes when ensemble averaging is removed.

* Corresponding author: wmoerner@stanford.edu

Looking back on 28 years of cryogenic single-molecule experiments

Michel Orrit*

Leiden University, LION, Huygens-Kamerlingh Onnes Lab, 2333CA Leiden, Netherlands

Abstract. Starting with the first single-molecule fluorescence experiments in 1990, the field of cryogenic single-molecule spectroscopy exploits the narrow zero-phonon lines of single molecules, usually in molecular crystals and glasses. Occasionally, similar experiments can also be done at room temperature, as illustrated by the case of the NV^- center in diamond. In this review contribution, I shall illustrate the variety and scope of the experiments performed in the past 28 years, highlighting some important points and outlooks.

1 Spectroscopy and magnetic resonance

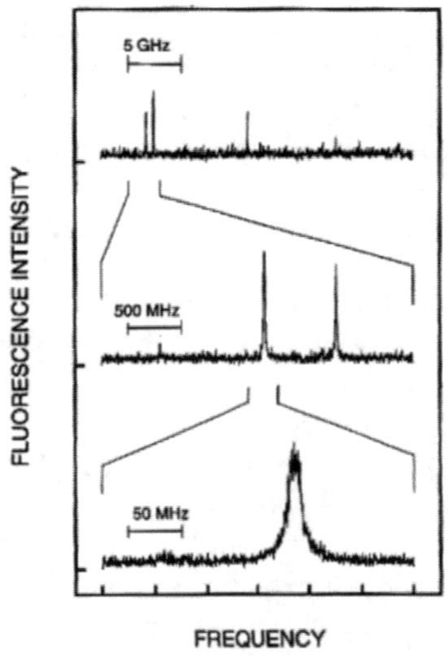

Fig. 1. Example of a single-molecule fluorescence excitation spectrum showing the lifetime-limited lines of single pentacene molecules in a p-terphenyl crystal at 1.8 K (from Ref. 1).

From the first high-resolution optical experiments on single molecules [1], it was clear that this fine spectroscopy enables accurate mapping of molecular electronic and vibronic states through excitation spectra and molecular vibrations from fluorescence spectra. Blinking single-molecule fluorescence yields information about triplet population and depopulation rates, and optically detected magnetic resonance provides the positions of the triplet sublevels split by magnetic spin-spin and spin-orbit interactions [2]. Later work identified hyperfine interactions and achieved magnetic resonance experiments of single nuclear spins through ENDOR [3].

---

* Corresponding author: orrit@physics.leidenuniv.nl

2 Quantum optics

The narrow zero-phonon lines are often lifetime-limited, which opens unique opportunities for quantum optics. Antibunching in the emission of single molecules was demonstrated early on at low temperature [4]. Single-photon emission at low temperature [5] provides indistinguishable photons, that have been used in Sandoghdar's group for pioneering quantum optics experiments in solid-state phases for the first time, before the experiments with self-assembled quantum dots. Today, single molecules at low temperature are still attractive contenders for single-photon sources delivering indistinguishable photons.

3 External field effects and local probing

Being ultra-narrow oscillators, single molecules are very sensitive probes for all kinds of perturbations, by pressure or mechanical strain, electric fields, magnetic fields. This sensitivity can be exploited to detect small changes in the surroundings of the molecules, for example from acoustic or mechanical deformation [6] or from charge carriers diffusing or trapped in the matrix.

4 Spectral diffusion

The time-dependent dynamics of single-molecule lines has clarified the concepts of spectral diffusion in molecular solids, glasses and polymers at low temperature, by displaying individual two-level systems, and quasi-local vibration modes [7]. Many other causes of spectral diffusion have been identified or proposed in more specific systems, such as phenyl ring flips in p-terphenyl crystals, or rotational-translational coupling in methylated matrixes.

In conclusion, the knowledge and experience accumulated in the previous two decades have made single organic molecules uniquely attractive quantum emitters. Along with self-assemble quantum dots and color centers in diamond, these fascinating systems are excellent candidates for a broad variety of nanophysics experiments, including optomechanics, localized probing, or as the cores of devices for the storage and treatment of quantum information.

References

1. M. Orrit and J. Bernard, Phys. Rev. Lett. **65**, 2716 (1990)
2. J. Wrachtrup, C. von Borczyskowski, J. Bernard, M. Orrit and R. Brown, Nature **363**, 244 (1993)
3. J. Wrachtrup, A. Gruber, C. von Borczyskowski, Chem. Phys. Lett. **267**, 179 (1997)
4. T. Basché, W. E. Moerner, M. Orrit, H. Talon, Phys. Rev. Lett. **69**, 1516 (1992)
5. C. Brunel, B. Lounis, P. Tamarat, M. Orrit, Phys. Rev. Lett. **81**, 2679 (1999)
6. Y. Tian, P. Navarro, M. Orrit, Phys. Rev. Lett. **113**, 135505 (2014)
7. A. V. Naumov, Y. G. Vainer, L. Kador, Phys. Rev. B **79**, 132201 (2009)

Diamond based quantum technologies

Fedor Jelezko[1,2,*]

[1]Institute of Quantum Optics, Ulm University, Germany
[2]Center for Integrated Quantum Science and Technology (IQst), Ulm University, Germany

Abstract. Diamond is not only the king gemstone, but also a promising material in quantum technologies. Optically active impurities (colour centers) in diamond show unique coherence properties under ambient conditions. Their quantum state can be readout and manipulated using a combination of single molecule spectroscopy and magnetic resonance techniques. In this talk it will be shown how engineered spins in diamond can be used for creation of non-classical (entangled) quantum states. I will also demonstrate the potential of atomic magnetometers based on single color centers for nanoscale sensing and imaging. New photoelectric detection technique allowing efficient readout of single color centers will be discussed.

Electron and nuclear spin resonance is a spectroscopic technique that provides very detailed information on almost every possible substance but also as very insensitive. This may be contrasted to the excellent sensitivity of optical single molecule spectroscopy which is able to demonstrate the detection of single molecules in condensed matter. A recent development is a combination of these techniques making it possible to detect magnetic resonance transitions of single quantum systems. While first experiments were done with aromatic molecules at low temperatures, recently single spin magnetic resonance techniques were also applied to the study of defect centers in diamond allowing coherent single spin control and readout under ambient conditions. This new approach opened a number of interesting applications, such as quantum information processing and communication, secure quantum communication and sensing with nanometer resolution.

A particularly interesting application of diamond based quantum sensing is the detection of nuclear magnetic resonance on nanometer scales, including the detection of individual nuclear spins or small ensembles of external nuclear spins. Single nitrogen vacancy (NV) color centers in diamond currently have sufficient sensitivity for detecting single external nuclear spins and resolve their position within a few angstroms. The ability to bring the sensor close to biomolecules by implantation of single NV centers and attachment of proteins to the surface of diamond enabled the first proof of principle demonstration of proteins labeled by paramagnetic markers and label-free detection of the signal from a single protein. Single-molecule nuclear magnetic resonance (NMR) experiments open the way towards unraveling dynamics and structure of single biomolecules. However, for that purpose, NV magnetometers must reach spectral resolutions comparable to that of conventional solution state NMR. New techniques were proposed for this purpose and

[*] Corresponding author: fedor.jelezko@uni-ulm.de

realized recently including technique that employs quantum entanglement. The ability to sense nuclear spins by NV centers also enables the transfer of polarization from optically polarized spins of NV centers to external nuclear spins. Such diamond based techniques for dynamic nuclear spin polarization are very promising for the enhancement of sensitivity of conventional MRI imaging.

Most of mentioned above results obtained so far with diamond centers are based on optical detection of single NV color centers. Recently it was shown that photoelectrical detection of NV centers base on spin selective photoionization can provide robust and efficient access to spin state of individual color centers.

Longitudinal current in single-atom interacting with laser field

Anatoli Andreev[1], *Sergei* Stremoukhov[1,2,*], and *Olga* Shoutova[1]

[1]Faculty of Physics, Lomonosov Moscow State University, 119991, Moscow, Russia
[2]National Research Centre "Kurchatov Institute", 123182, Moscow, Russia

The progress of the scientific research in the last century shows the following: the macroscopic phenomena and devices based on them find soon their realizations in microphysics. That is the case of the one-atom lasing device, the one-electron transistor, etc. Here, we discuss another phenomenon of such kind, the longitudinal current in a single atom [1,2]. Such generalization of the longitudinal wave concept is absolutely self–consistent. The atomic current density is strongly defined in material wave theory. By calculating the atomic current density we can divide it into the transversal and longitudinal parts which are orthogonal and parallel to the driving laser field wave vector, respectively. The appearance of the longitudinal waves or currents is usually associated with structured fields, i.e. tightly focused beams, evanescent or interference fields. In the case of single atom the appearance of the longitudinal component in atomic current is due to the breaking of spherical symmetry when the atomic electron interacts with superposition of intra-atomic field and the field of external wave. Indeed the superposition of intra-atomic field and field of the external plane wave ceases to be the spherically symmetric. The non-perturbative theory of light-atom interaction [3] enables us to calculate the response field polarization and what is more important the dependency of polarization of emitted waves on the polarization of the driving laser field.

The effect of the high order harmonic generation (HHG) in the intense laser fields is under permanent study for the three last decades. During this long period a number of different theoretical models have been proposed (see for review [3]). Most commonly the three-step model is used. This model is more semi-intuitive than self-consistent, as it combines the quantum process of ionization, the classical evolution of ionized electron in the field of the laser pulse, and finally the recombination of accelerated electron on the parent ion. The common feature of these models consists in a supposition that the HHG is a process of inelastic laser field scattering by atom, molecule, ion, etc. In analogy with the early models of the second or third harmonic generation it is assumed that the energy of a harmonic is a sum of energies of laser radiation quanta. For example, the three quantum process $\omega_3 = n_1\omega_1 + n_2\omega_2$ in femtosecond two-colour field $\omega_1 = \omega$ and $\omega_2 = 2\omega$ results in fourth harmonic generation at $n_1 = 2,\ n_2 = 1$ or in terahertz radiation generation at $n_1 = 2,\ n_2 = -1$, etc.

The alternative approach to the theory of light-atom interaction was proposed in [3]. This method is based on the use the new complete orthonormal basis for expansion of the time-dependent Schrodinger equation wave-function. The proposed basis consists of time-dependent eigenfunctions of boundary value problem for "an atom in the external field" [3]. Here, the HHG can be interpreted as a parametric process, and the energy of emitted harmonics results from the work produced by laser field on electron movement in the intra-atomic field. In the frames of this theory, the HHG in the double-frequency field is a process when the generation of N-th harmonic is due to the interaction of the following type $N\omega = n_1\omega + n_2 \cdot 2\omega$, where $n_{1,2}$ can be both positive and negative.

In the frames of dipole approximation, it is assumed that the linearly polarized light interacts with z component of dipole moment while circularly polarized light interacts with $(x \pm iy)$ components

* Corresponding author: sustrem@gmail.com

of atomic electron radius vector [5]. It should be noted that the orientation of the atomic configuration space relates to the angular momentum orientation: the quantum-mechanical average of the angular momentum $\langle nlm|\vec{l}\,|nlm\rangle = \vec{n}_z\hbar m$ is oriented along the z axis of atomic configurational space.

Here, we present the specific feature of the single atom response which can be consistently interpreted in the frames of the theory [3]: the appearance of the atomic current longitudinal component which is orthogonal to the polarization vector of the incident laser field. The longitudinal component appears even in the case of a single-color laser field. The effect opens a way towards the development of new methods of the generated radiation polarization control.

The origin of the longitudinal atomic current is in following. Indeed, the spatial distribution of the atomic electron density $\rho(\vec{r},t) = e|\psi(\vec{r},t)|^2$ is related to the atomic current density through the continuity equation $\partial\rho(\vec{r},t)/\partial t + \operatorname{div}\vec{j}(\vec{r},t) = 0$. In the free atom case, when the orbital quantum number of ground state $l \geq 1$, the electron density distribution is cylindrically symmetric with respect to z axis of atomic configuration space. The direction of this axis coincides with the direction of the quantum-mechanical average $<\vec{l}>$ of the atomic angular momentum. In the external laser field, the distribution of the electron density is influenced by the polarization of the driving laser field. Let us consider a spherically symmetric distribution of the electron density in s state. In the external laser field the electron density distribution takes the cylindrical symmetric form with an axis directed along the laser field polarization vector. The elongation of the electron density distribution in one direction obviously results in a contraction of the distribution radius in the perpendicular plane. This effect leads directly to the appearance of the longitudinal current or, more precisely, of the components of the atomic current perpendicular to the direction of the laser field polarization vector. It is evident that in the case of $l = 0$ the integral longitudinal current is equal to zero due to the cylindrical symmetry of the electron density distribution in the external field. However, if $l > 0$ and the angular momentum and field strength are not collinear, then the cylindrical symmetry is broken.

The results demonstrate clearly that the longitudinal component of the atomic current is efficiently induced when an atom with a non-zero angular momentum interacts with plane or one multicolour electromagnetic wave. This component is directed along the wave vector of the incident field and its amplitude depends on the geometry of the interaction, i.e. on the angles between the atomic angular momentum and the polarization vector (and wave vector) of the incident wave.

The spectrum of the longitudinal atomic current includes both odd and even harmonics of the incident field. The ratio of the amplitudes of odd and even harmonics depends on the orientation of the atomic angular momentum with respect to the polarization vector and wave vector of the driving laser field. So, we can easily control the spectrum of the HHG by varying the mutual orientation of the laser field polarization vector and the atomic angular momentum.

The account of the longitudinal component opens ways to control the spectral structure of the atomic current. It might be critical for the polarization state of the generated radiation. Usually, the detector system lies in line with the laser propagation direction [6], and the longitudinal projection of the current affect the final ellipticity only as an additional "dimension" which take energy away from the other projections. As a result, one can control the energy swop for the given harmonics. However, the impact of the longitudinal component can be more significant if the detector system has a non-zero angle with the laser field propagation direction. In that case one should take into account the longitudinal projection in ellipticity calculations. Account of the longitudinal projection can be critical for a consistent explanation of the results obtained in non-collinear geometries [7].

The work was partially supported by the RFBR under projects Nos. 18-02-00528, 18-02-00743, 18-02-40014.

References

1. A. Andreev, S. Stremoukhov, O. Shoutova, EPL, **120**, 14003 (2017)
2. A. Andreev, S. Stremoukhov, O. Shoutova, JETP, **127**, 25 (2018)
3. A. Andreev, S. Stremoukhov, O. Shoutova, Eur. Phys. J. D, **66**, 16 (2012)
4. P. Corkum, Phys. Rev. Lett., **171**, 1994 (1993)
5. L.D. Landau, E.M. Lifshitz, *Quantum Mechanics* (Pergamon Press, London-Paris, 1958)
6. S. Stremoukhov, A. Andreev, B. Vodungbo et al., Phys. Rev. A, **94**, 013855 (2016)
7. M. Louisy, C. L. Arnold, M. Miranda et al., Optica, **2**, 563, (2015)

Imaging and force transduction in correlative scanning force and confocal fluorescence microscopy

Thomas Basché[*]

Institute of Physical Chemistry, Johannes Gutenberg-University, Mainz, Germany

Abstract. Correlative scanning force and confocal fluorescence microscopy has been used to study individual molecules, nanoparticles and nanoparticle oligomers. By applying a compressive force via the AFM cantilever, spectral blue and red shifts in the range of several meV/GPa have been observed for single dye molecules and semiconductor quantum dots. Moreover, individual Au nanoparticle dimers linked by a chlorophyll binding protein have been imaged in both modes and plasmonic fluorescence enhancement of the chlorophyll emission of up to a factor of 15 has been found.

Correlative microscopies are gaining increasing attention, since they allow for correlating structure or morphology of individual nano-sized objects from a potentially heterogeneous population with their electronic or photo-physical properties. An early example have been measurements of the emission spectra and polarization of single semiconductor nanocrystal quantum dots as well as their crystallographic structure by a combination of confocal fluorescence microscopy and transmission electron microscopy [1]. Recently, relying on the same approach it has been shown that the fraction of "on"-times of single CdSe/CdS nanocrystals decreased when stacking faults were present [2]. Besides transmission electron microscopy, single molecule spectroscopy has also been combined with scanning electron microcopy or scanning probe techniques [3].

We have used confocal fluorescence and scanning force (AFM) microscopy of single molecules and nanoparticles for two purposes: i) To correlate structural information with photo-physical properties, and ii) To apply compressive stress and track the effects of the impact on the electronic states by fluorescence spectroscopy.

In a first set of experiments we applied compressive forces via the AFM cantilever to single terrylenediimide molecules decorated with large side-groups and deposited on a mica surface. Reversible blue and red-shifts as well as irreversible shifts were observed and attributed to different conformations of the terrylene core induced by the compressive force [4]. Similar experiments were conducted with CdSe/CdS/ZnS nanocrystals and again blue and red shifts were registered under the directional force [5]. These experiments clearly have shown that a given particle either shifts to the red or blue under pressure (see Fig.1), resolving a controversial issue remaining from bulk studies of semiconductor quantum dots

[*] Corresponding author: thomas.basche@uni-mainz.de

in diamond anvil cells. The direction of the spectral shifts largely depends on the crystallographic axis of the quantum dot onto which the non-hydrostatic pressure is applied. However, the orientation of the approximately spherical core-shell particles typically is not known. This obstacle can be mediated by the use of nanocrystal platelets which can be deposited in a well-defined way and therefore expose a known crystallographic face.

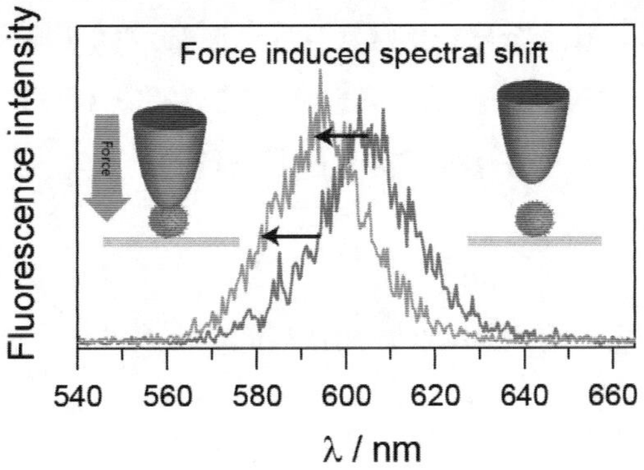

Fig. 1. Scheme how to apply force to a single nanoparticle with the tip of an AFM-cantilever. (after Ref. 5) For the case shown the emission spectrum of a single CdSe/CdS/ZnS-nanocrystal was shifted to the blue with increasing force. After force was released, the spectrum shifted back to its original position.

Correlative atomic force and confocal microscopy has also been employed to study the properties of Au nanoparticle oligomers, linked by the water-soluble chlorophyll binding protein (WSCP). For individual dimers - as identified by AFM - fluorescence enhancements of up to a factor of 15 with a clear dependence on the excitation polarization were observed and attributed to mainly plasmonic enhancement of the *Chl* excitation rate. Promising features of this protein-directed assembly of a plasmonic nanoantenna are the simple transfer of the construct into different environments and the structure-driven positioning of the pigments in the plasmonic hotspot.

References

1. F. Koberling, U. Kolb, G. Philipp, I. Potapova, Th. Basché, A. Mews, J. Phys. Chem. B **107**, 7463 (2003)
2. N.J. Orfield, J.R. McBride, J.D. Keene, L.M. Davis, S.J. Rosenthal, ACS Nano **9**, 831 (2015)
3. M. Vacha, D. Sharma, S. Hirata, J. Photochem. Photobiol. C **34**, 121 (2018)
4. S. Stöttinger, G. Hinze, G. Diezemann, I. Oesterling, K. Müllen, Th. Basché, Nature Nanotechnology **9**, 182 (2014)
5. T. Fischer, S. Stöttinger, G. Hinze, A. Bottin, N. Hu, Th. Basché, Nano Lett. **17**, 1559 (2017)

Correlation between the maximum wavelength and the line width in the emission of a single semiconductor colloidal quantum dot at different temperatures

Eduard Podshivaylov[1], *Maria* Kniazeva[1], *Ivan* Eremchev[2], *Andrey* Naumov[2,3], and *Pavel* Frantsuzov[1,*]

[1]Lomonosov Moscow State University, 119991 Moscow, Russia
[2]Institute of Spectroscopy RAS, 108840 Moscow, Russia
[3]Moscow State Pedagogical University, 119435 Moscow, Russia

The very first measurements [1] of the spectrally resolved photoluminescence of single colloidal semiconductor quantum dots CdSe (QDs) revealed the phenomenon of very slow spectral diffusion (SD) with a characteristic time scale in order of hundred seconds. Müller et al. [2,3] found correlation between linewidth and peak position of the single QD spectrum at temperatures from 5 K to 300 K. Usual explanation of this phenomenon [2-4] is based on the quantum-confined Stark effect by the movement of charge carriers around individual QDs. If this hypothesis is correct, the effect should depend on the dielectric permittivity of the media. Gomez et al. [5] found no dependence on the dielectric permittivity.

A unique spectroscopic experiment on single QD under slow heating and cooling was carried out by our group. Fluorescence images and spectra of single quantum dots were measured via home-build fluorescence microscope equipped by prism spectrometer. Two optical schemes: wide-field and scanning confocal schemes were combined in one microscope in order to simplify procedure of single quantum dots preliminary searching (by fluorescence image processing and antibunching determination) and sequential measuring of fluorescence spectra of the selected QD.

Quantum dots (CdSe/ZnS, Sigma Aldrich with Fluorescence peak at 620 nm) were dispersed in toluene solution of polyisobutylene in low concentration and then spincoated onto a cover glass. Thicknesses of polymer films with single quantum dots were varied in the range of several tens of nanometers. Sample temperature was controlled by LakeShore temperature controller and regulated by Peltier element attached to the sample. The experiment presented here was carried out at ambient conditions.

Tunable dye laser (Coherent CR599) was used to excite quantum dots at the wavelength of 580 nm near the quantum dot absorption band edge. Excitation laser intensity (~ 100 W/cm^2 in a focused spot) was attenuated by neutral spectral density filters (Standa) and controlled by Newport Power meter. Set of interference filters separated fluorescence

* Corresponding author: pavel.frantsuzov@gmail.com

signal from scattered laser radiation. Two highly sensitive cooled EMCCD cameras were utilized to capture single quantum dot images and spectra. Hanbury Brown and Twiss scheme with broadband 50% splitter and two identical SPAD detectors with time resolution 1 ns and dead time 200 ns (QE ~ 50 %) were used to measure autocorrelation function for QD fluorescence intensity. Some details of the apparatus and measurements procedures were described previously in [6,7].

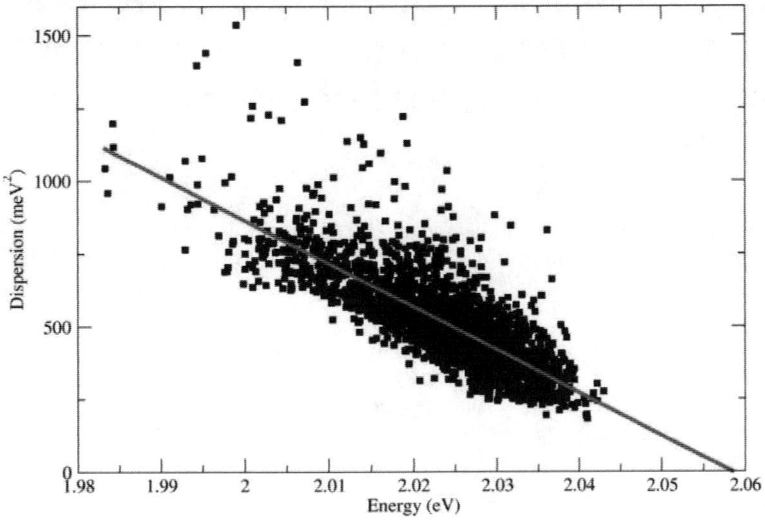

Fig. 1. The dependence of the dispersion on the peak wavelength. The points correspond to the experimentally obtained values at room temperature. The solid line indicates the theoretical dependence obtained in the proposed model based on the fluctuations of the electron-phonon interaction.

Fluorescence spectra sequences for single quantum dots were measured with exposure time 200 ms and spectral resolution ~ 0.7 nm, which was enough to rich good signal to noise ratio.

The experiments show a correlation between the position of the maximum and width of the QD emission spectrum (see Fig. 1). In order to explain experimental results we propose a new model of the QD spectral diffusion and temperature dependent shift. In the model both the peak position and the linewidth are determined by the electron-phonon interaction value. Slow variations of this value in time lead to correlated fluctuations of the wavelength and linewidth at given temperature.

The study was supported by the Russian Foundation for Basic Research, project 16-02-00713.

References

1. S. A. Empedocles, D. J. Norris, and M. G. Bawendi, Phys. Rev. Lett. **77**, 3873 (1996)
2. S. A. Empedocles and M. G. Bawendi, Science, **278**, 2114 (1997)
3. J. Muller, J. M. Lupton, A. L. Rogach, J. Feldmann, D.V. Talapin, and H. Weller, Phys. Rev. Lett., **93**, 167402 (2004)
4. J. Muller, J. M. Lupton, A. L. Rogach, J. Feldmann, D.V. Talapin, and H. Weller, Phys. Rev. B, **72**, 205339 (2005)
5. D. E. Gómez, J. van Embden, and P. Mulvaney, Appl. Phys. Lett., **88**, 154106 (2006)
6. I. Y. Eremchev, I. S. Osad'ko, A. V. Naumov, J. Phys. Chem. C, **120**, 22004 (2016)
7. I. S. Osad'ko, I. Y. Eremchev, A. V. Naumov, J. Phys. Chem. C, **119**, 22646 (2015)

EPJ Web of Conferences **190**, 02004 (2018) https://doi.org/10.1051/epjconf/201819002004
HBSM-2018

Quantum photonic integrated circuits with waveguide integrated superconducting nanowire single-photon detectors

Gregory Goltsman[1,2,*]

[1]Department of Physics, Moscow State Pedagogical University, Moscow 119992, Russia
[2]Zavoisky Physical-Technical Institute of the Russian Academy of Sciences, Kazan 420029, Russia

Abstract. We show the design, a history of development as well as the most successful and promising approaches for QPICs realization based on hybrid nanophotonic-superconducting devices, where one of the key elements of such a circuit is a waveguide integrated superconducting single-photon detector (WSSPD). The potential of integration with fluorescent molecules is discussed also.

Currently, many physical systems for the realization of qubits including superconducting circuits, trapped ions and atoms, quantum dots, color centers in a solid and photons are intensely studied. Here we focus only on the last item in the list - photons. Because photons interact only weakly with an optically transparent medium but not with each other, have several degrees of freedom for encoding of quantum information and provide fast propagation speed, they are an attractive choice for creating quantum networks. However, while individual photons can be manipulated conveniently, the creation of two- and three- deterministic qubit gates requires strong nonlinear interaction between photons. In this case, the advantages of the photons are compensated for by their shortcomings, which for a long time limited the use of photons as qubits. Fortunately, Knill, Laflamme and Milburn (KLM) proposed the concept of linear optical quantum computing (LOQC), allowing to create non-deterministic gates using photons, linear optical elements and single photon detectors. While the implementation of the KLM-protocol is possible in free space, the need for a large number of optical components and their precise configuration requires a more complex solutions. Due to a number of advantages, such as scalability, small footprint, inherent stability, no need for optical alignment as well as low power consumption and CMOS-compatibility, quantum photonic integrated circuits (QPICs) can successfully overcome these challenges.

Popular material platforms for the realization of QPICs include silicon, gallium arsenide, and diamond. Each platform has its own advantages and disadvantages and is currently in different stages of development [1]. Nevertheless, all of the platforms rely on combining single photon sources, linear optical elements and single-photon detectors on chip.

* Corresponding author: goltsman@rplab.ru

Recently, a fully integrated circuit including a single photon source (carbon nanotube), detectors (superconducting nanowire single-photon detectors) and silicon-nitride waveguides has already been implemented on a single chip [2]. Despite the demonstration of the working concept, the purity of the emulsion of electroluminescent photons is not yet sufficient for large scale processing and new single-photon sources are constantly being investigated.

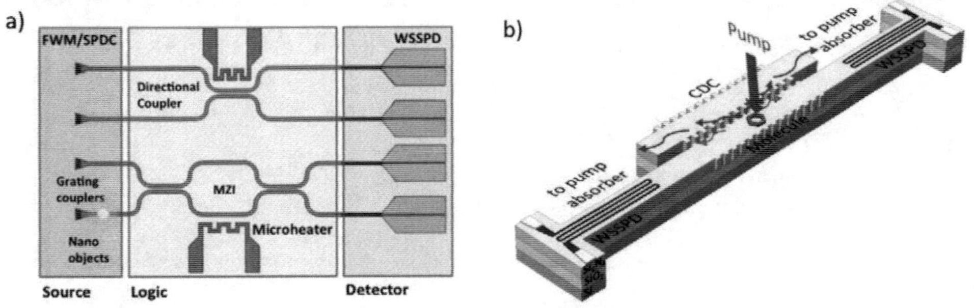

Fig. 1. a) Schematic view of the platform for the realization of QPIC. The nanophotonic chip includes single-photon sources, single photon detectors and quantum logic. FWM is four wave mixing and SPDC is spontaneous parametric down-conversion. b) Schematic view of a nanophotonic circuit with on-chip fluorescent single molecule

Fluorescent molecules are one of the promising single-photon sources. Recently the emission of single photons from Dibenzoterrylene (DBT) molecules into Si_3N_4 ridge waveguide at room temperature was demonstrated. DBT was placed on a Si_3N_4 waveguide to couple emitted single photons through the evanescent waveguide mode into the guided mode. The advantages of this approach include a small footprint, simple fabrication methods and scalability towards arrays of integrated single photon sources for QPICs. However, unlike the on-chip integrated electroluminescent nanotube [3], a more complex photonic circuit is required. The next step in the integration of a single-photon emitted molecule and QPIC including effective pump rejection (CDC) and superconducting nanowire single-photon detectors (WSSPDs) is proposed in Fig. 1b. As an efficient on-chip filter a contra-directional coupler (CDC) can be used. After optical pumping of a single molecule, the residual pump is rejected to the neighboring waveguide and routes to the optical absorber. Emitted photons are directed to WSSPD, reproducing the Hanbury Brown-Twiss scheme. By measuring the on-chip second order correlation function, one can detect non-classical light for QPICs, as well as to study the molecules itself.

Going beyond such proof-of-principle concepts, the realization of large scale QPICs is expected to have profound impact on science and technology, material engineering, as well as quantum information processing including quantum computing, simulation and metrology.

G. Goltsman acknowledges support by Ministry of Education and Science of the Russian Federation № 14.583.21.0065.

References

1. S. Bogdanov, M. Y. Shalaginov, A. Boltasseva, A. Shalaev. Opt. Mater. Express **7**, 111 (2016)
2. S. Khasminskaya, F. Pyatkov, K. Słowik, S. Ferrari, O. Kahl, V. Kovalyuk, P. Rath, A. Vetter, F. Hennrich, M. M. Kappes, G. Gol'tsman, A. Korneev, C. Rockstuhl, R. Krupke, and W. H. P. Pernice. Nat. Photonics **10**, 727 (2016)

Impact of single point mutations on the excitonic structure and dynamics in FMO complex

Anton Khmelnitskiy[1] and *Ryszard* Jankowiak[1,2]

[1]Department of Chemistry, Kansas State University, Manhattan, KS, USA
[2]Department of Physics, Kansas State University, Manhattan, KS, USA

The Fenna–Matthews–Olson (FMO) antenna complex from the green sulfur bacterium *Chlorobaculum (C.) tepidum* is an important model protein to study exciton dynamics and excitation energy transfer in photosynthetic complexes. Single point mutations provide an excellent opportunity for the study of exciton structure and interexciton level relaxation. In this bacterium, the energy harvested by chlorosomes is funnelled via the baseplate pigments and FMO trimers to the reaction center (RC), where photochemistry takes place. The structure of the FMO protein from *C. tepidum* was solved and revealed a homotrimeric complex containing eight bacteriochlorophylls (BChl) *a* per monomer [1].

In this work we focus on two mutants: 1) V152N (here mutation occurs near BChl 1, i.e. valine is replaced by asparagine); and 2) W184F (where tryptophan 184 near BChl 6 is replaced with phenylalanine). In order to obtain information on the excited state lifetimes (T_1) and interexciton level relaxation in the mutants we measured low temperature absorption/emission spectra and performed high-resolution resonant HB experiments. It is well known that the excited state lifetimes, that are related to the so-called homogeneously broadened lines (Γ_{hom}), are easily obtained via hole-burning spectroscopy as $1/T_2 = (\pi c\Gamma_{hom})^{-1} = (\pi T_1)^{-1} + (\pi T_2^*)^{-1}$, where T_2 is the total dephasing time and T_2^* is the pure dephasing time; c is the velocity of light. At T = 4 K Γ_{hom}, obtained for excitonically coupled systems are strongly dominated by the exciton level decay times (T_1). $T_1 = (2\pi c\Gamma_{hom})^{-1}$, where $\Gamma_{hom} = \frac{1}{2}\,\Gamma_{ZPH}$. In the modeling studies we apply a recently developed algorithm [2] to simultaneously fit multiple optical spectra. Intra-monomer interactions are modeled using non-Markovian Redfield theory. Calculations are performed for the FMO trimer.

Fig. 1 compares the extracted resonant holes (four black narrow spikes) obtained for the typically studied wild type (WT) FMO with holes burned in the V152N mutant (Frame A; ten red holes) and W16F mutant (Frame C; eleven red holes). Spectra a (black) and b (red) are the absorption spectra obtained at 4 K for typical WT FMO, V152N (Frame A) and W16F (Frame C) mutant samples respectively. All resonant holes in Fig. 1 are obtained with a constant fluence f = 6 J/cm^2.

The absence of resonant bleach in WT FMO sample at $\nu_B \geq 12450$ cm^{-1}, as illustrated in Fig. 1A/B, implies very fast (femtosecond) energy relaxation in this spectral range in agreement with ultrafast pump-probe and 2DES data [3]. That is, the lifetimes of the states excited in this region are too short to allow for an observable resonant bleach (broad ZPH). Likewise, the presence of ZPHs at higher energies in V152N (Fig. 1A) and W16F (Fig. 1C) indicates to slower (ps) energy relaxation for higher energy excitons compared to WT FMO. The slower experimentally estimated T_1 values for V152N (~1 ps) and W184F (1.2-1.9 ps) mutants for high energy excitons are also in good agreement with the averaged calculated values of 1.1 and 1.3 ps, respectively. Figure 2 shows experimental/calculated absorbance (a/a′), emission (b/b′) and nonresonant hole-burned (c/c′) spectra for V152N mutant.

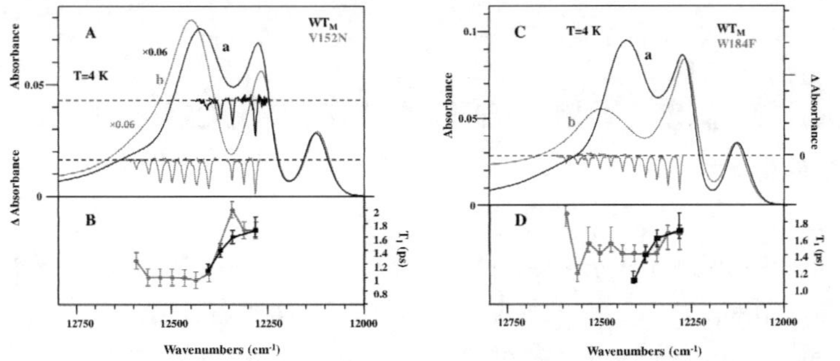

Fig. 1. Spectra a and b are the 4 K absorption spectra obtained for a WT FMO (black curve a), V152N mutant (Frame A: red; curve b) and W184F mutant (Frame C: red; curve b). The top four holes (Frame A: black curves) and the red holes (at the bottom) correspond to the extracted resonant holes observed in a typical WT FMO, V152N mutant (Frame A) and W184F mutant (Frame C) respectively. The excited-state decay times (T_1) calculated from hole widths (Γ_{ZPH}) obtained as a function of burn frequency (ν_B) for WT (black squares), V152N mutant (Frame B: red circles) and W16F mutant (Frame D: red squares) with corresponding error values (bars).

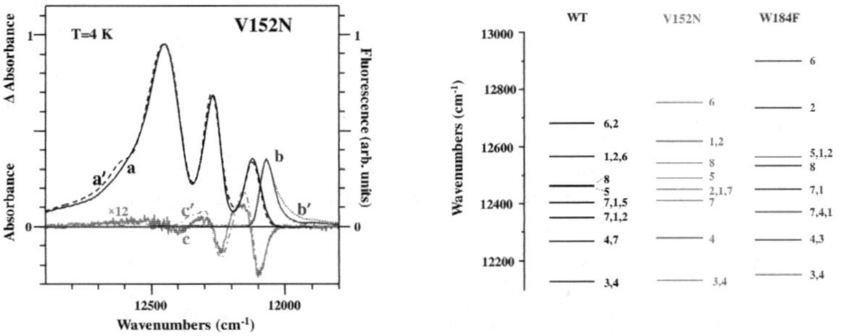

Fig. 2. Left frame: Black spectra are the experimental/simulated (solid/dashed curves a/a′). Blue curves show the experimental/simulated (solid/dashed curve b/b′). Red curves c and c′ correspond to the experimental (solid) and calculated (dashed-dotted) nonresonant HB spectra; ν_{ex} = 20,490 cm^{-1} for V152N FMO mutant at T = 4 K. Right frame: exciton energy level scheme for monomers of WT and three mutants based on our calculations. Dominant pigment contributions (>15%) for each exciton are indicated by corresponding BChl numbers.

Figure 2 (right frame) shows calculated exciton energy level scheme for WT and the two mutants. It is evident that mutations near BChl 1 and 6 result in red shift of higher energy excitons. Thus, we suggest that the site-energy shifts of the BChls 1 and 6 in particular, shared by several high-energy excitons, are responsible for the observed changes in higher energy excited-state decay times. As a result, those states shift to higher energies and overlap between them decreases. Moreover, longer excited-state decay (T_1) times of highest energy excitons in V152N and W184F mutants suggest that site energies of BChls 1 and 6, believed to play an important role in receiving excitation from the baseplate BChls, likely play a critical role to ensure the fs energy relaxation observed in wild-type FMO complexes.

References

1. J. Wen, H. Zhang, M. L. Gross, R. E. Blankenship, Biochemistry, **50**, 3502 (2011)
2. A. Khmelnitskiy et al., Biochim. Biophys. Acta, **1859**, 165 (2018)
3. E. Thyrhaug et al., J. Phys. Chem. Lett., **7**, 1653 (2016)

EPJ Web of Conferences **190**, 02006 (2018) https://doi.org/10.1051/epjconf/201819002006
HBSM-2018

Light controls light: single molecules as optical switches

Johannes Maier[1], *Martti* Pärs[1], *Katja* Gräf[2], *Mukundan* Thelakkat[2], and *Jürgen* Köhler[1,*]

[1]Spectroscopy of soft Matter, University of Bayreuth, 95447 Bayreuth, Germany
[2]Applied Functional Polymers, University of Bayreuth, 95447 Bayreuth Germany

Abstract. In recent years much attention has been given to design multistate molecular components with functionalities that cover the range from simple switches to logic gates [1-3]. In this regard photochromic molecules, i.e., molecules that can be interconverted between two bistable forms by light, have played an important role. Promising candidates that fulfill obvious demands such as high photochemical/ photophysical stability and high fatigue resistance are compounds of the family of diarylethenes [2,3]. However, a serious drawback of this class of molecules is a low fluorescence quantum yield. Therefore we adapted the strategy developed by Irie and coworkers [2,3], to chemically synthesize complex tailor-made triads consisting of a photochromic dithienylcyclopentene (DCP) unit covalently linked to two peryline bisimide (PBI) molecules that are known as strong fluorophores, see fig.1 inset top left. This facilitates the combination of high fatigue resistance and high fluorescence quantum yield. Illumination with light in the UV spectral region induces a ring-closure reaction of the DCP and leads to a state with suppressed fluorescence from the PBIs, whereas light in the VIS spectral region yields a ring opening reaction of the DCP and restores the fluorescence from the PBI units. This allowed us to verify functionalities like optical gating and amplifying, yet where the electrons have been replaced by photons as signal carriers [4-6], see fig.1.

Switching of photochromic molecules at the single molecule level has been demonstrated before [7], yet in those experiments no distinction was made between deliberate switching and unavoidable blinking of single molecules. The switching process is manifested as a finite delay between the onset of the fluorescence and the onset of the VIS laser. For a sequence of 8800 switching events the distribution of delay times can be reasonably well approximated as an exponential decay, in agreement with the expectation for a switching process that obeys Poissonian statistics [8]. From a statistical analysis, that takes results from reference experiments performed on the same triad without switching into account, we determined the positive predictive value p(S|+) for deliberate light-induced switching. In words: The conditional probability that the triad indeed has switched given that we observed a finite delay for the onset of the fluorescence. For 15 triads studied the average of this probability amounts to 0.8 ± 0.1. In the best cases it reached about 96%,

* Corresponding author: juergen.koehler@uni-bayreuth.de

insufficient for optical memories with single triads, yet exceptionally good for super-resolution microscopy.

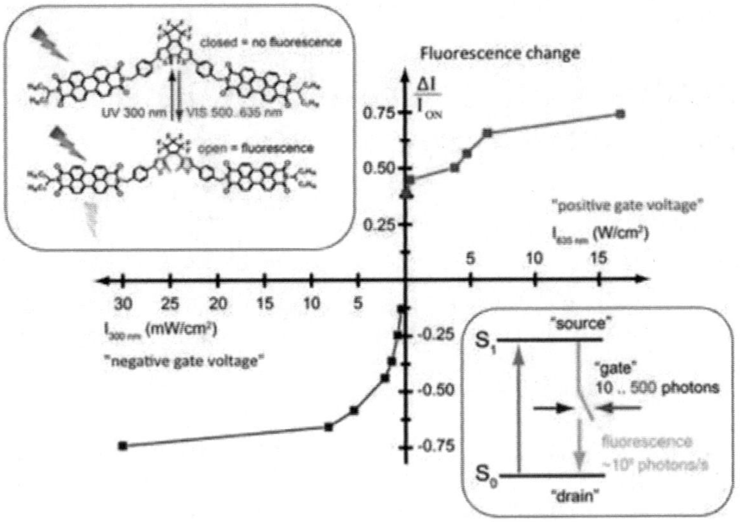

Fig. 1. Relative change of the fluorescence of the photochromic triads as a function of the intensity of the UV/VIS conversion beams. The flow of (many) fluorescence photons from the PBI units can be controlled by (few) UV and VIS photons that induce the photocyclization / photocycloreversion reactions of the photochromic DCP unit, respectively.

References

1. B. L. Feringa, *Molecular Switches* (Wiley-VCH, Weinheim, 2001)
2. M. Irie, Chem. Rev., **100**, 1685 (2000)
3. M. Irie, T. Fukaminato, K. Matsuda, S. Kobatake, Chem. Rev. **114**, 12174 (2014)
4. M. Pärs, C. C. Hofmann, K. Willinger, P. Bauer, M. Thelakkat, J. Köhler, Ang. Chem. Int. Ed., **50**, 11405 (2011)
5. M. Pärs, K. Gräf, P. Bauer, M. Thelakkat, J. Köhler, Ap. Phys. Lett., **103**, 221115 (2013)
6. M. Pärs, M. Gradmann, K. Gräf, P. Bauer, M. Thelakkat, J. Köhler, Scientific Reports, **4**, 4316 (2014)
7. M. Irie, T. Fukaminato, T. Sasaki, N. Tamai, T. Kawai, Nature, **420**, 759 (2002)
8. J. Maier, M. Pärs, T. Weller, M. Thelakkat, J. Köhler, Scientific Reports, **7**, 41739 (2017)

EPJ Web of Conferences **190**, 02007 (2018) https://doi.org/10.1051/epjconf/201819002007
HBSM-2018

λ/2 Fabry Pérot micro-resonators in single molecule spectroscopy

Alfred J. Meixner[1,*], *Frank* Wackenhut[1], *Alexander* Konrad[1], *Michael* Metzger[1], and *Marc* Brecht[1,2]

[1]Eberhard Karls University, Institute of Physical and Theoretical Chemistry, 72076 Tübingen, Germany
[2]Reutlingen University, Reutlingen Research Institute, Process Analysis and Technology (PA&T), 72762 Reutlingen, Germany

Abstract. Embedded in a tuneable λ/2-FabryPérot micro-resonator the radiative relaxation of a dye molecule or quantum dot can reproducibly be modified allowing to determine their quantum yield, control Förster energy-transfer or localize them with nanometer precision.

1 Introduction

Small optical micro-resonators are structures that confine light to volumes with dimensions on the order of a wavelength and hence have become increasingly important for controlling and studying light-matter interaction in integrated optics [1]. The spontaneous emission rate of an atom or molecule depends on both the transition dipole moment (TDM) between the electronically excited state and the electronic ground state as well as on the density of optical modes of the electromagnetic field in the surrounding medium. In an optical subwavelength micro-resonator with piezo-controlled mirror separation the density of optical modes can be varied with respect to the free space in a controlled manner, and hence, also the radiative coupling of the dipole transition of the emitter to the optical field can be tuned.

2 Results

We demonstrate controlled modification of the vibronic relaxation dynamics of a single fluorescent molecule fixed in a solid host matrix by varying the distance between the cavity mirrors. While it is well known from quantum electrodynamics nowadays that the spontaneous emission is not an intrinsic property of an atom but can be modified by tailoring the electromagnetic environment, the situation for a molecule or quantum dot embedded in a solid host is more complex since the radiation rate and relaxation pathways depend on the balance between the radiative (the far-field) and non-radiative (the near-field) relaxation dynamics, the latter being sensitively dependent on the interaction with the local chemical environment. Hence, by controlling the electromagnetic environment of a single molecule or quantum dot, we can select the vibronic transition where fluorescence will mostly occur [2] and we can tune its radiative transition and determine its fluorescence quantum yield [3] as shown in Fig. 1).

* Corresponding author: alfred.meixner@uni-tuebingen.de

Using a tuneable optical micro-resonator with subwavelength spacing, we demonstrate controlled modulation of the radiative transition rate of a single molecule, which is measured by monitoring its fluorescence lifetime [4]. By comparing the experimental data with a theoretical model, we extract both the pure radiative transition rate as well as the quantum yields of individual molecules. We observe a broad scattering of quantum yield values from molecule to molecule, which reflects the strong variation of the local interaction of the observed molecules with their host environment. Finally, we will also discuss the possibility of controlling Förster resonance energy transfer in a tuneable micro-resonator [5, 6, 7] and nanometer precise localization [8].

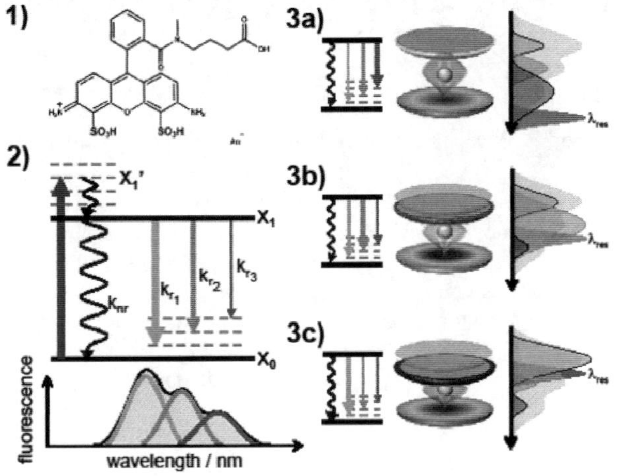

Fig. 1. Adopted from ref. [5] (1) Chemical structure of a typical dye molecule. (2) Jablonski-like diagram with an electronic ground state X_0 and excited state X_1. After pulsed excitation (violet arrow) into the excited vibronic level X_1' the system relaxes to the purely electronic level X_1. The electronic relaxation occurs either non-radiatively or radiatively into possible vibronic levels of the electronic ground state. Each fluorescent transition can be described by a spectrally broadened Gaussian band due to thermal energy level fluctuations and phonon interactions with the environment. (3a–3c) Tuning the micro-resonator with the on-axis spectral emission to the emission wavelength of a specific vibronic transition enhances resonant and suppresses off-resonant radiative transitions.

References

1. H. Yokoyama, K. Ujihara, *Spontaneous emission and laser oscillation in microcavities* (CRC Press, 1995)
2. A. I. Chizhik, F. Schleifenbaum, R. Gutbrod, A. M. Chizhik, D. Khoptyar, A. J. Meixner, J. Enderlein, J. Phys. Rev. Lett., **102**, 073002 (2009)
3. A. I. Chizhik, A. M. Chizhik, D. Khoptyar, S. Bär, A. J. Meixner, J. Enderlein, Nano Lett., **11**, 1700 (2011)
4. A. I. Chizhik, A. M. Chizhik, A. M. Kern, T. Schmidt, K. Potrick, F. Huisken, Phys. Rev. Lett., **109**, 223902 (2012)
5. F. Schleifenbaum, A. M. Kern, A. Konrad, A. J. Meixner, et al. PCCP, 16, 12812 (2014)
6. A. Konrad, M. Metzger, A. M. Kern, M. Brecht, A. J. Meixner, Nanoscale, **7**, 10204 (2015)
7. A. Konrad, A. L. Trost, S. Skandary, M. Hussels, A.J. Meixner,N. V. Karapetyan, M. Brecht, Phys. Chem. Chem. Phys., **16**, 6175 (2014)
8. R. Gutbrod, D. Khoptyar, M. Steiner, A. M. Chizhik, A. I. Chizhik, S. Bär, A. J. Meixner, Nano Lett., **10**, 504 (2010)

Investigation of evaporation of sessile droplets using luminescent nano-probes and other applications of NV centers in diamond

Taras Plakhotnik* and *Haroon* Aman

School of Mathematics and Physics, The University of Queensland, St Lucia, QLD 4072, Australia

Abstract. The paper describes application of diamond nano crystals to research on dynamic processes in small (less than 1 mm across) evaporating droplets deposited on a solid substrate. Such droplets are used as a model system for testing proposed bio applications of nitrogen-vacancy centers in diamond. We demonstrate that a high spatial resolution of our methods reveals unexpected features of the evaporation and fluid mechanics in such droplets.

Research on nitrogen-vacancy centers in diamond (NV-centers) has gained its momentum about 30 years ago and since then the field has developed in several directions. One direction of research mainly exploits very stable luminescence of these centers, which may in bulk crystal environment can escape photo bleaching during several weeks of continues photo excitation. If such centers are embedded in very small (tens of nanometers or less) crystals, these crystals can be used as luminescent labels in bio imaging applications (such as tracking and targeted visualization) where they successfully compete with conventionally used organic dyes and quantum dots. However the necessity to use nano-crystals has a number of undesired consequences of practical importance such as significantly reduced luminescence rate [1].

In this report we demonstrate applications of nano-diamonds for investigation of fluid dynamics in small evaporating droplets deposited on a solid surface. This system (where physics is not yet completely described with quantitative models) is interesting by itself but also provides a simple playground for developing and testing concepts employed in corresponding intracellular applications.

A very popular physical model for the evaporating droplet has been suggested 20 years ago in a seminal paper [2] and it predicts that the velocity of the fluid in the evaporating droplet deposited on a solid substrate increases towards the rim of the droplet. This velocity is proportional to $x^{-\lambda}$, where x is the distance to the rim while the value of $\lambda = (\pi-2\theta)/(2\pi-2\theta)$ depends on θ, the contact angle between the droplet and the substrate.

The experimental results obtained by tracing a 35-nm diamond with about 20-nm accuracy are shown in Fig. 1. Contrary to the theoretical predictions, the data do not show acceleration towards the rim and careful analysis of diffusion points to a decrease of velocity of liquid within less than 2-μm distance from the rim.

* Corresponding author: taras@physics.uq.edu.au

A very interesting prospect for NV centers is their sensitivity to various external parameters such as magnetic and electrical field and temperature [3]. Magnetic field sensors based on NV-centers exploit a not zero spin of the electronic ground state of these defects in diamond lattice and the dependence of the luminesce intensity on the spin state (the value of its projection on the center axis). Temperature measurements can either use the sensitivity of the ODMR frequency to the crystal temperature (the resonance shifts with a gradient of 75 kHz/K [4]) or the thermal sensitivity of the zero-phonon line (ZPL) in the luminescence spectra (ZPL of NV-centers is clearly visible even at room temperature) [5].

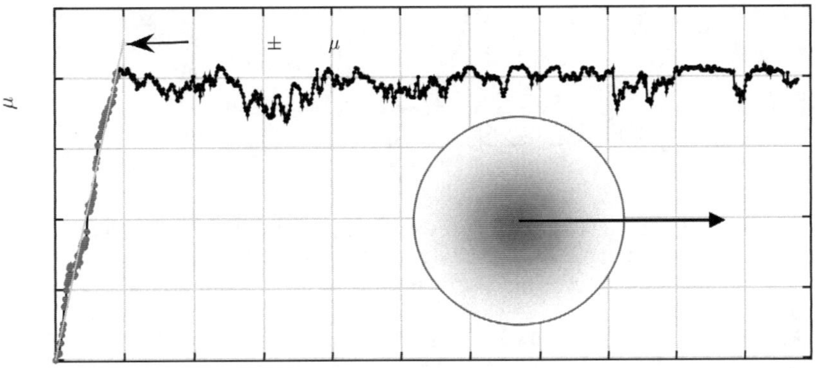

Fig. 1. The *x*-position vs. time for a nano-diamond when it approaches the rim of a droplet. The initial position on the *x*-axis and the time are set to zero at the start of tracing. The position of the rim corresponds to the displacement $x \approx 21$ μm. For ≈ 2 s the diamond drifts towards the rim with a constant velocity of about 11.4 μm/s and then its movement is described near the rim as constrained diffusion in a steady liquid flow. The inset shows the droplet and the direction of the *x*-axis.

Temperature measurements inside biological cells have raised recently a hot debate among physicist, biophysicists and biologists. The discussion is focused on the temperature rise in single cells under physiological conditions. Simple physical arguments suggest that a cell generate very small temperature gradients (even on a 100-nm scale the estimated temperature difference is less than 1 mK). But some experimental results show a local temperature rise on the order of 1 K. The contradiction has been even named as "The 10^5 gap" paradox [6]. Most of the measurements in this area are done using thermo sensitive dyes and their reliability is questioned due to their sensitivity to other factors. The solution of this contradiction requires new methods for sensitive and ultra local temperature measurements. Because most of the reported measurements of temperature in evaporating droplets also lack reliability, we propose to investigate evaporative cooling in micro-droplets using NV-centers in nano-crystals as thermometers. Such a model system is a reasonable proxy to the complex intracellular measurements.

References

1. T. Plakhotnik, H. Aman, Diamond Rel. Mater. **82**, 87 (2018)
2. R. D. Deegan et al, Nature **389,** 827 (1997)
3. T. Plakhotnik, Curr. Opin. Solid State Mat. Sci. **21**, 25 (2017)
4. V. M. Acosta et al. Phys. Rev. Lett. **104**, 070801 (2010)
5. T. Plakhotnik, H. Aman, H.-C. Chang, Nanotechnology **26**, 245501 (2015)
6. M. Suzuki et al., Nature Methods **12,** 802 (2015)

EPJ Web of Conferences **190**, 02009 (2018) https://doi.org/10.1051/epjconf/201819002009
HBSM-2018

Multiphoton spectroscopy: An optical window into molecular electrostatics

Aleksander Rebane[1,2,*]

[1]Montana State University, Physics Department, Bozeman, Montana, 59717, USA
[2]National Institute of Chemical Physics and Biophysics, Tallinn, 12618, Estonia

Abstract. Quantitative knowledge about static molecular electric dipole moments is essential for understanding of intramolecular charge transfer as well as nanometer-scale static electric interactions. However, measuring or determining the molecular electrostatic properties with sufficient accuracy remains a challenging task. In our experiments, we measure the femtosecond two-photon absorption spectra- and cross sections of a range of organic- and organometallic chromophores in solution and use these data to determine the electric dipole moment change in corresponding lowest-energy dipole-allowed transition. Good correspondence of our experimental dipole moments with the quantum-chemical calculations as well as reports by other groups using conventional dipole moment measurement methods suggests that quantitative multiphoton spectroscopy may offer all-optical alternative to the traditional techniques such as Stark effect and electrochromism.

Two-photon absorption (2PA), first proposed by Maria Goeppert-Mayer [1], consists in an instantaneous 3rd–order nonlinear-optical process, where two simultaneously incident photons induce a transition from molecule's ground state to an excited electronic state. Under ambient illumination conditions, the probability of 2-photon transitions as well as higher-order multiphoton absorption (MPA), remains notoriously low. However, recent proliferation of high peak intensity, wavelength-tuneable femtosecond lasers, make 2PA and MPA conveniently accessible in many laboratories, rendering these effect particularly useful for microscopy and biological imaging. Furthermore, recent progress towards higher accuracy and reliability of nonlinear spectroscopic measurements [2], have transformed 2PA into a precision quantitative method, with increasing range of applications.

Two-photon spectroscopy is able to take advantage of the empirical fact that if, in a dipolar chromophore, the lowest-energy 1-photon electronic transition is dipole-allowed, then the 1PA and 2PA spectral profiles of to the same purely electronic transitions, have coinciding shapes, and that the corresponding 2PA cross section is proportional to the square of the permanent electric dipole moment change in that transition [3]. This approach was recently used to evaluate static dipole moments of some dipolar dyes in solution and also estimate the strength of local dielectric reaction field [4]. On the other hand, if a molecular system maintains inversion symmetry, then according to Laporte spectroscopic selection rule, all permanent electric dipole moments vanish, and the 2PA transitions between opposite parity states are forbidden. However, under certain conditions, e.g. if the molecule undergoes

* Corresponding author: arebane@montana.edu

steric distortions or is placed within a complex nano-environment, then actual ground state symmetries may be lowered. In this case, experimental comparison between 1-photon and 2-photon absorption spectra may also yields useful insights regarding spontaneously induced permanent electric dipole moment change [5].

We describe our recent experiments, where we apply 2-photon excitation with a wavelength-tuneable femtosecond laser to measure the 2PA spectra with high accuracy in a broad range of wavelengths of a variety of molecular systems, including organometallic charge transfer complexes [6-8], fluorescent proteins [9], as well as DNA nucleotides [10]. In particular, ferrocene complexes show in their UV-vis absorption spectra prominent features associated with metal-to-ligand charge transfer (MLCT) transitions, where a notable fraction of the electron density shifts from the Fe atom(s) in the ground electronic state towards the organic group(s) in an excited state. We used our novel 2PA spectroscopy approach to quantify the corresponding large change of the molecular permanent electric dipole moment, both in non-centrosymmetric as well as nominally symmetric ferrocene oligomer structures, whereas in the latter case spontaneous symmetry breaking was attributed to ground state steric distortions [7]. Most recently, we used accurate measurement of the 2PA spectrum of ruthenium(II) tris-complexes of 2,2'-bipyridine and 1,10-phenanthroline ligands to determine the molecular electric dipole change in the corresponding MLCT transitions [8]. Where applicable, our experiments show good correspondence with reports by other groups using conventional dipole moment measurement methods such as Stark effect and electrochromism. We will also compare our experimental findings with quantum-chemical theoretical calculations and discuss how to optimize the calculated 2PA spectra and dipole moment values by choosing appropriate methods [11]. In summary, our results indicate that our 2-photon absorption based approach may be a viable alternative to more traditional methods of determining the electric dipole change based, while offering higher versatility associated with all-optical access.

References

1. M. Goeppert-Mayer, Annalen der Physik, **9**, 273 (1931)
2. S. de Reguardati, J. Pahapill, A. Mikhailov, Y. Stepanenko, and A. Rebane, Optics Express, **24**, 9053 (2016)
3. A. Rebane, N. S. Makarov, M. Drobizhev, B. Spangler, E. S. Tarter, B. D. Reeves, Ch. W. Spangler, F. Meng, and Z. Suo, J. Phys. Chem. C., **112**, 7997 (2008)
4. A. Rebane, G. Wicks, M. Drobizhev, T. Cooper, A. Trummal, and M. Uudsemaa, Angew. Chem. Int. Ed., **54**, 7582 (2015)
5. A. Rebane, M. Drobizhev, N. S. Makarov, G. Wicks, G. P. Wnuk, Y. Stepanenko, J. E. Haley, D. M. Krein, J. L. Fore, A. R. Burke, J. E. Slagle, D. G. McLean, and T. M. Cooper, J. Phys. Chem. A, **118**, 3749 (2014)
6. T. M. Cooper, J. E. Haley, D. M. Krein, A. R. Burke, J. E. Slagle, A. Mikhailov, and A. Rebane, J. Phys. Chem. A, **121**, 5442 (2017)
7. A. Mikhaylov, M. Uudsemaa, A. Trummal, E. Arias, I. Moggio, R. Ziolo, T.M. Cooper, and A. Rebane, J. of Phys. Chem. Lett., **9**, 1893 (2018)
8. C. Stark, M. Rammo, J. Pahapill, K. Petrisenko, A. Mikhaylov, and A. Rebane, Proc. SPIE, **10529**, 105291D (2018)
9. M. Drobizhev, N. S. Makarov, S. E. Tillo, T.E. Hughes, and A. Rebane, Nature Methods, **8**, 393 (2011)
10. A. Mikhaylov, S. de Reguardati, J. Pahapill, P. R. Callis, B. Kohler, A. Rebane, Biomed. Opt. Expr., **9**, 447 (2018)
11. M. Uudsemaa, A. Trummal, S. de Reguardati, P. Callis, A. Rebane, Phys. Chem. Chem. Phys., **19**, 28824 (2017)

EPJ Web of Conferences **190**, 02010 (2018) https://doi.org/10.1051/epjconf/201819002010
HBSM-2018

Organic molecules coming of age in quantum optics

Vahid Sandoghdar*

Max Planck Institute for the Science of Light, Staudtstraße 2, 91058 Erlangen, Germany

The interaction of light and matter at the nanometer scale lies at the heart of quantum optics because it concerns elementary processes such as absorption or emission of a photon by an atom. Over the past decade, we have shown that direct coupling of a photon to a single two-level atom should be possible via tight focusing. However, because transitions in quantum emitters are typically not closed, laboratory demonstrations of this idea fall short of the theoretical prediction. In this presentation I shall report on recent achievements, where the branching ratio of a single organic molecule is improved by a substantial Purcell effect when coupled to a microcavity. Furthermore, we will discuss coherent linear and nonlinear experiments on molecules coupled to subwavelength waveguides on a chip. Together with their ability to generate narrowband stream of single photons, these developments make organic molecules viable candidates for integration in chip-based quantum optical circuits.

* Corresponding author: vahid.sandoghdar@mpl.mpg.de

EPJ Web of Conferences **190**, 02011 (2018) https://doi.org/10.1051/epjconf/201819002011
HBSM-2018

Non-radiative processes in metal halide perovskite semiconductors probed by photoluminescence microscopy

Aboma Merdasa[4], *Marina* Gerhard[1], *Boris* Louis[1,3], *Jun* Li[1], *Alexander* Dobrovolsky[1], *Yuxi* Tian[3], *Johan* Hofkens[3], *Rafael* Camacho[3], *Eva* Unger[2], and *Ivan G.* Scheblykin[1,*]

[1]Chemical Physics and NanoLund, Lund University, PO Box 124, SE-22100, Lund, Sweden
[2]School of Chem. & Chem. Engineering, Nanjing University, 22 Hankou Rd, Nanjing 210023, China
[3]Department of Chemistry, KU Leuven, Celestijenlaan 200F, B-3001 Leuven, Belgium
[4]Helmholtz-Zentrum Berlin für Mater. und Energie GmbH, Kekulestraße 5, 12489 Berlin, Germany

Abstract. Organo metal halide perovskites are solution processed semiconductors that recently attracted a great attention. They possess a rather "soft" and (photo) chemically active solid structure allowing for ion migration and other mass diffusion processes. This is a likely reason why non-radiative recombination centres in these materials are activated and de-activated on relatively slow time-scales. This dynamics reveals as photoluminescence (PL) fluctuations (blinking) of individual microcrystals and local areas of films and allows for application of a broad range of single molecule spectroscopy methods including optical super-resolution. Studying PL blinking resolves properties of individual non-radiative centres and helps to unravel their chemical nature.

Non-radiative charge recombination in semiconductors sets a fundamental limitation to the efficiencies of light emitting devices and solar cells. Recently re-discovered organo-metal-halide perovskites (OMHP) have become the most popular and, in a way, mysterious optical semiconductors showing a great potential for industrial applications. It was quickly realized that spectroscopy methods able to monitor the fate of photogenerated charge carriers locally with high spatial resolution are of a high demand. [1] The resolution in space is needed because these solution-processed semiconductors possess highly inhomogeneous properties in terms of polycrystallinity, defect state distributions, formation of intermediates and degradation. Photoluminescence (PL) micro-spectroscopy is one of such methods. OMHPs possess phenomena of photo-brightening, defect curing, ion migration, and other light-induced photochemical processes which can be nicely monitored by PL micro-spectroscopy [1]. In several publications we have successfully used ideas coming from single molecule spectroscopy to study perovskite films and individual nano- and micro-crystals [2-4]. OMHP materials are highly dynamic in terms of local chemical structure leading to easily observable temporal fluctuations of PL yield of localized regions of perovskite films [5] or individual crystals [2-4]. PL flickering or blinking is the direct indication of switching processes in efficient non-radiative recombination centres.

Non-radiative recombination in semiconductors is a difficult process to comprehend in

[*] Corresponding author: ivan.scheblykin@chemphys.lu.se

detail [6,7]. It occurs due to the presence of intra-band defect states (traps) or complex non-radiative centres. The concentration of these quenchers and the charge trapping cross sections per one quencher are usually not known. Luckily OMHPs possess properties that allowed us to see and count each individual non-radiative channel in a small particle. These properties are crystal structure dynamics and mass (ion) diffusion, leading to transformation of the material at slow time scale of milliseconds, seconds and beyond. It turned out that these dynamics lead to activation/deactivation of individual non-radiative channels observed experimentally as PL blinking. The efficiency of just one channel can be very large, allowing for on/off switching of the PL of 500 nm sized $MAPbI_3$ crystals and substantial fluctuations of the PL of 10 micrometer long $MAPbI_3$ rods with a volume of about 1 μm^3 [2,4]. It was suggested that such an efficient quencher (a super-trap or a "killing" centre) can be a donor-acceptor complex [2]. Monitoring of PL fluctuations allows us to obtain information about charge carrier and exciton diffusion, trapping and trap concentrations. Moreover, the switching allows for application of optical super resolution spectroscopy to localize trapping and emitting sites in space with nanometer accuracy [2,8].

Single molecule fluorescence and its blinking gave birth to super-resolution optical microscopy (Nobel Prize 2014 in chemistry). Similarly, fluctuating activity of individual quenchers in OMHPs allowed us to get a mechanistic view beyond ensemble averaging on the non-radiative recombination in these semiconductors. We were able to resolve the temperature-dependent action of each defect state individually by observing 50 nm sized $MAPbI_3$ crystals and studying the increase of their PL intensity upon cooling from 300 K to 77 K. We suggest that the temperature activation of PL quenching is related to the potential barriers present in charge carrier diffusion and in charge trapping process. Contrary to some literature [9] we propose that the apparent activation energy of the PL intensity is not directly related to the exciton binding energy. Individual crystals showed highly diverse temperature dependencies of PL, which in many cases could not be approximated by the Arrhenius equation. Further studies are on the way towards Monte-Carlo simulations of the PL blinking dynamics and estimation of activation energies of the non-radiative centre formation and passivation and correlating these energies with the activation barriers of mass diffusion processes in perovskites including ion migration.

References

1. C. S. Ponseca, Y. Tian, V. Sundström, and I. G. Scheblykin, Nanotechnology **27**, 82001 (2016)
2. A. Merdasa, Y. Tian, R. Camacho, A. Dobrovolsky, E. Debroye, E. L. Unger, J. Hofkens, V. Sundström, and I. G. Scheblykin, ACS Nano **11**, 5391 (2017)
3. A. Dobrovolsky, A. Merdasa, E. L. Unger, A. Yartsev, and I. G. Scheblykin, Nat. Commun. **8**, 34 (2017)
4. Y. Tian, A. Merdasa, M. Peter, M. Abdellah, K. Zheng, C. S. Ponseca, T. Pullerits, A. Yartsev, V. Sundström, and I. G. Scheblykin, Nano Lett. **15**, 1603 (2015)
5. A. Halder, R. Chulliyil, A. S. Subbiah, T. Khan, S. Chattoraj, A. Chowdhury, and S. K. Sarkar, J. Phys. Chem. Lett. **6**, 3483 (2015)
6. V. N. Abakumov, V. I. Perel, and I. N. Yassievich, *Nonradiative Recombination in Semiconductors* (North-Holland, Amsterdam, 1991)
7. S. D. Stranks, ACS Energy Lett. **2**, 1515 (2017)
8. Y. Tian, A. Merdasa, E. Unger, M. Abdellah, K. Zheng, S. McKibbin, A. Mikkelsen, T. Pullerits, A. Yartsev, V. Sundström, and I. G. Scheblykin, J. Phys. Chem. Lett. **6**, 4171 (2015)
9. T. J. Savenije, C. S. Ponseca, L. Kunneman, M. Abdellah, K. Zheng, Y. Tian, Q. Zhu, S. E. Canton, I. G. Scheblykin, T. Pullerits, A. Yartsev, and V. Sundström, J. Phys. Chem. Lett. **5**, 2189 (2014)

EPJ Web of Conferences **190**, 02012 (2018) https://doi.org/10.1051/epjconf/201819002012
HBSM-2018

Photophysics and electroluminescence of single nanocrystals of halide perovskites and related nanomaterials

Martin Vacha*, and *Dharmendar Kumar* Sharma

Department of Materials Science and Engineering, Tokyo Institute of Technology, Ookayama 2-12-1, Meguro-ku, Tokyo 152-8552, Japan

Abstract. We report simultaneous photoluminescence and electroluminescence single-particle study of nanocrystals of inorganic halide perovskite $CsPbBr_3$, as well as of ternary I-III-IV semiconductor quantum dots.

Due to their exceptional excitonic properties such as strong absorption and efficient narrow-band emission, lead halide perovskite nanocrystals (NCs) have been recently rediscovered as possible alternatives to semiconductor quantum dots for electroluminescence (EL) devices and other light-emitting and sensing applications, and research into their basic photophysical properties is attracting increasing attention. While dispersions of NCs generally show high photoluminescence (PL) quantum yields, the efficiencies of electrically driven luminescence devices are considerably lower. We have shown previously that simultaneous characterization of PL and EL on single-molecule level can bring valuable insight into these emission processes in conjugated polymers [1] and small molecules [2]. Especially, the differences in luminescence blinking and spectral dynamics between the two processes can provide information on the different dynamics of the excited states and their relaxations. Here, we report a simultaneous study of EL and PL of inorganic halide perovskite nanocrystals with the aim to uncover the origin and ways of suppression of blinking in both EL and PL to maximize the emission quantum yield, as well as to study the origin of spectral diffusion and spectral shifts, in relationship with structural/compositional stability.

We use nanocrystals of $CsPbBr_3$ prepared by methods reported in literature. In solution, the NCs show efficient and narrow emission in the green part of the spectrum. We first examined the influence of a surrounding matrix on their spectral properties and blinking behavior by dispersing the NCs in an inert matrix of PMMA and in a conductive matrix of PVK/PBD, both on the level of single NCs and aggregates (arrays) of tens of NCs. The differences in the blinking characteristics appear as different parameters of truncated power law distributions of the on-times, both for bright and dim (grey) states. These findings are further corroborated by an effect of externally applied electric field on the blinking dynamics and PL spectra. The applied field enhances the blinking in both matrices, and

* Corresponding author: vacha.m.aa@m.titech.ac.jp

causes a quadratic Stark shift, the extent of which differs from particle to particle. Based on the above characterization, we next fabricate a single-particle organic light-emitting device (OLED) by dispersing the NCs at very low concentrations in PVK/PBD matrix as an emitting layer in an ITO/PEDOT:PSS/PVK/PBD/TPBi/LiF/Al device and study the device by PL and EL single-particle spectroscopy. As shown in an example in the *Figure 1*, the same NCs can exhibit dramatically different brightness and blinking behavior upon the different modes of excitations. Such differences in PL and EL characteristics can be attributed partly to influence of charge injection and trapping in the device, but intrinsic properties of the NC arrays also play a role in the observed phenomena.

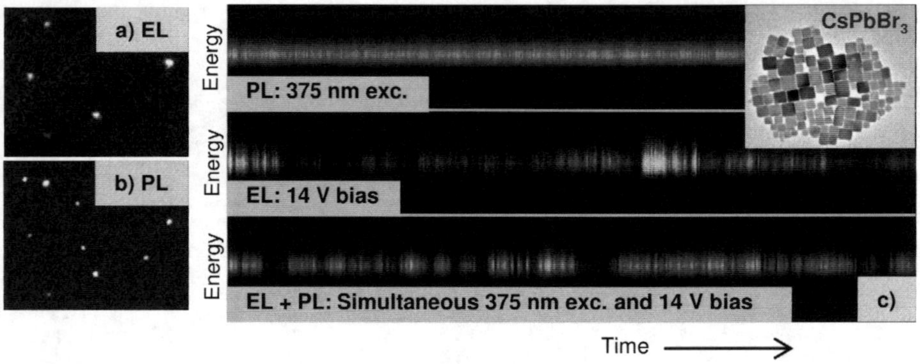

Fig. 1. Electroluminecsnece (a) and photoluminescence (b) microscopic images of $CsPbBr_3$ nanocrystals in an ITO/PEDOT:PSS/PVK/PBD/TPBi/LiF/Al device. (c) Spectral time traces measured from one bright spot in the images upon excitiation by light, electric field and combined show different blinking and lineshape dynamics.

Apart from the work on perovskite nanocrystals, we report on recent progress in single-particle photophysical characterization of ternary I-III-IV semiconductor nanocrystals which have been explored as a non-toxic alternative to II-VI semiconductor quantum dots. For NCs of $(AgIn)_xZn_{2(1-x)}S_2$ we explore the origin of defect emission, multiple emitting sites and suppression of blinking [3], and report the emergence of narrow band-edge PL in core/shell structures [4].

References

1. Y. Honmou, S. Hirata, H. Komiyama, J. Hiyoshi, S. Kawauchi, T. Iyoda and M. Vacha, Nature Commun. **5**, 4666 (2014)
2. Y. Sekiguchi, S. Habuchi and M. Vacha, ChemPhysChem **10**, 1195 (2009)
3. D. K. Sharma, S. Hirata, L. Bujak, V. Biju, T. Kameyama, M. Kishi, T. Torimoto and M. Vacha, Nanoscale **8**, 13687 (2016)
4. T. Uematsu, K. Wajima, D. K. Sharma, S. Hirata, T. Yamamoto, T. Kameyama, M. Vacha, T. Torimoto, S. Kuwabata, NPG Asia Mater. DOI : 10.1038/s41427-018-0067-9 (2018)

Upconversion nanoparticles: on the way from diagnostics to theranostics

Alla Generalova[1,2,*], *Kristina* Mironova[1,2], *Natalya* Sholina[3], *Vasylina* Rocheva[2], *Andrey* Nechaev[4], *Ekaterina* Grebenik[3,5], *Anna* Guller[5], *Andrey* Zvyagin[5], *Sergey* Deyev[1], *Vitaly* Zubov[1], and *Evgeny* Khaydukov[2,3]

[1]Shemyakin-Ovchinnikov Institute of Bioorganic Chemistry RAS, 117997, Moscow, Russia
[2]Scientific Research Centre "Crystallography and Photonics" RAS, 119333, Moscow, Russia
[3]Sechenov First Moscow State Medical University, 119991 Moscow, Russia
[4]Institute of Fine Chemical Technologies, 119571, Moscow Russia
[5]Macquarie University, Sydney, NSW 2109, Australia

Abstract. We report of surface modification approaches of nanoparticles with anti-Stokes luminescence, known as upconversion nanoparicles (UCNPs), comprised of inorganic host $NaYF_4$ codoped with Yb^{3+} and Er^{3+}or Tm^{3+}. These approaches enabled the facile, lossless preparation of hybrid polymer-encapsulated UCNPs suitable for bioassays. These probes inherited UCNP properties, such as excellent photoluminescence under excitation with NIR light from the biotissue "transparency window", as well as they were dispersible in aqueous media and physiological buffers, exhibiting chemical stability. The feasibility of the hybrid UCNPs was demonstrated for *in vitro* bioassay and *in vivo* optical whole animal imaging using a home-built epi-luminescence imaging system.

1 Introduction

Lanthanide-doped upconversion nanoparticles (UCNPs) represents a flexible platform, where various surface moieties can be attached enabling targeting, bioimaging and therapy in a broad physiological context. UCNPs are characterized by sharp emission, high conversion efficiency, long lifetimes, low cytotoxicity, negligible photobleaching, and high spatial-temporal resolution during bioimaging. The efficient conversion of near-infrared (NIR) excitation at the wavelength of 980 nm into the shorter-wavelength IR, visible and UV spectral range emission (known as "upconversion") represents its most acclaimed property. Under the NIR excitation of UCNPs the excitation of tissue autofluorescence is almost negligible. Additionally, the excitation light penetration in biological tissue is greater in comparison with visible light, up to one centimeter. Generally, the UCNPs are based on efficient host matrix $NaYF_4$, which produces anti-Stokes luminescence, co-doped with lanthanide ions: Yb^{3+} as a sensitizer, and Er^{3+} or Tm^{3+} as activator.

* Corresponding author: a-generalova@yandex.ru

2 Results and Discussion

Typically, UCNPs are synthesized in organic solvents, resulting in UCNPs stabilized with hydrophobic oleic acid ligand. This fact limits their biological applications. Therefore, surface modification of UCNPs is a crucial step in an effort to create the water-dispersible and biocompatible UCNP probes.

Modification method using amphiphilic polymer poly(maleic anhydride-*alt*-1-octadecene) have been developed, which made it possible to create target-based UCNP hybrid nanocomplexes for *in vitro* labeling the HER2 / neu cancer-specific marker (Fig.1, way 1) [1]. We exploited the original design of UCNP nanocomplexes, where a "corona" was obtained from chemically bound PEG molecules, in order to increase the circulation time in the blood system *in vivo* and deliver UCNPs to the tumor (Fig.1, way 2) [2]. UCNP encapsulation into polyacrolein particles in the course of radical heterophase polymerization was the basis for production the reagents capable to study the particle biodistribution by organs at intravenous injection (Fig.1, way 3) [3]. Unique optical properties of UCNPs enabled formation of hybrid nanocomplexes with riboflavin (Rf) responsible for generating active oxygen forms under the NIR-radiation through the realization of FRET effect. Hybrid nanocomplexes were successfully applied for photodynamic therapy and showed tumor regression in mice, following the Rf-UCNPs peritumoural injection and near-infrared light photodynamic treatment of the lesions (Fig.1, way4) [4].

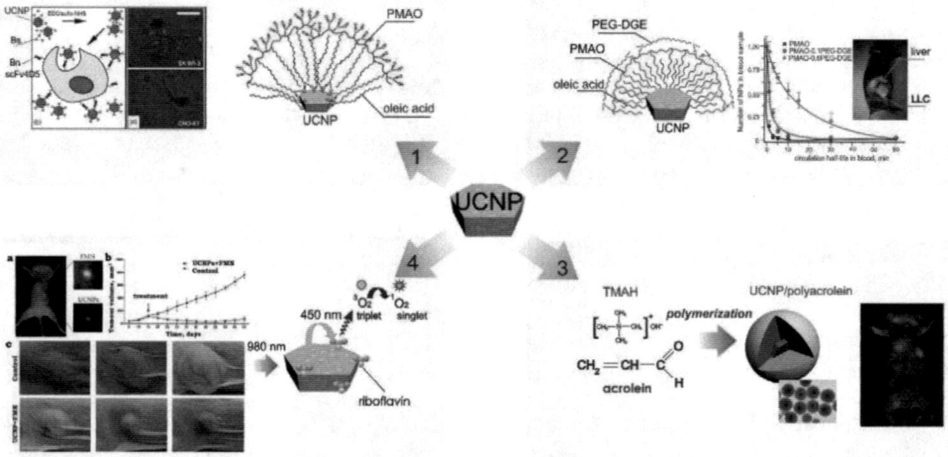

Fig. 1. Approaches of UCNP surface modification for various bioapplications.

This work was partially supported by Russian Foundation for Basic Research [RFBR research project №17-03-01033, RFBR research project № 17-00-00122 (K) (17-00-00118)]

References

1. E.A. Grebenik, A. Nadort, **A.N.** Generalova, et al., J. of Biomedical Optics, **18**, 076004 (2013)
2. A.N. Generalova, V.V. Rocheva, A.V. Nechaev, et al., RSC Advances, **6**, 30089 (2016)
3. A.N. Generalova, I.K. Kochneva, E.V. Khaydukov, et al., Nanoscale, **7**, 1709 (2015)
4. E.V. Khaydukov, K. E. Mironova, V.A. Semchishen, et al., Sci. Rep., **6**, 35103 (2016)

EPJ Web of Conferences **190**, 03002 (2018) https://doi.org/10.1051/epjconf/201819003002
HBSM-2018

Local and macroscopic characterization with single molecules and single quantum emitters

Andrei Naumov[1,2], *Alexey* Gorshelev[1], *Maxim* Gladush[1,2,*], *Jurgen* Köhler[3,4,5] and *Lothar* Kador[3]

[1]Institute of Spectroscopy, Russian Academy of Sciences, 108840 Troitsk, Moscow, Russia
[2]Moscow State Pedagogical University, 108840 Moscow, Russia
[3]University of Bayreuth, Institute of Physics, D-95440 Bayreuth, Germany
[4]University of Bayreuth, Soft-Matter Spectroscopy, D-95440 Bayreuth, Germany
[5]Bavarian Polymer Institute, D-95440 Bayreuth, Germany

Abstract. We introduce an experimental approach for mapping effective local values of dielectric characteristics of solid films and the analysis of the related local-field effects. The characterisation technique is based on the imaging and spectroscopy of single chromophore molecules at cryogenic temperatures. The progress in the theory of fluorescence enhancement due to the local field effects is reported.

We report the progress in development of the experimental approach for mapping effective values of local dielectric characteristics of frozen solid films and the analysis of the effects either enhanced or suppressed by the local fields. This probing technique is based on imaging and spectroscopy of single chromophore molecules at cryogenic temperatures. Since the fluorescence lifetime of a single dye molecule imbedded in a transparent matrix depends on the properties of the encapsulating medium at different spatial scales it has long been a challenge to detect inhomogeneities in the host material by simply measuring the lifetime of individual molecules distributed over the sample [1]. The factors that change the excitation lifetime in the matrix are known to be related to existence of the local density of the photon states and the local response of the medium to incident light. Both factors may be attributed to local effective values of the dielectric function describing the continuous surroundings of each chromophore light emitter. Thus, measuring the radiative lifetime distribution may help revealing the distribution of effective permittivities or refractive indices as well as the pattern of the local fields. In our experiment the spatial mapping of the local values is accomplished by localizing the corresponding chromophores with nanometer accuracy. We demonstrate this approach for a polycrystalline film doped with terrylene molecules [2]. We also report a significant progress in the theory of light emission enhancement due to the local field effects [3].

This work was partially supported by the Russian Foundation for Basic Research (RFBR № 17-02-00652).

* Corresponding author: mglad@isan.troitsk.ru

References

1. T. A. Anikushina, M. G. Gladush, A. A. Gorshelev, A. V. Naumov, Faraday Discussions, **184**, 263 (2015)
2. A. V. Naumov, A. A. Gorshelev, M. G. Gladush, T. A. Anikushina, A. V. Golovanova, J. Köhler, and L. Kador, Nano Letters, DOI: 10.1021/acs.nanolett.8b01753 (2018) – in press
3. D.V. Kuznetsov, Vl. K. Roerich, M. G. Gladush, Theor. Math. Phys., **168**, 1078 (2011)

Entangled plasmon generation in nonlinear spaser systems

Alexander Shesterikov[1], *Sergei* Karpov[1], and *Mikhail* Gubin[2,1]

[1]Vladimir State University named after A. G. and N. G. Stoletovs, Department of Physics and Applied Mathematics, 600000 Vladimir, Russia
[2]Moscow State Pedagogical University, Russian Institute for Advanced Studies, 119991 Moscow, Russia

Abstract. The work is focused on the investigation of features of quantum dynamics for localized plasmons in spaser systems consisting of metal nanoparticles (NP) and semiconductor quantum dots (QDs). The non-classical plasmon states generation in a three-particle spaser system with nonlinear plasmon-exciton interaction is predicted.

In this paper we propose a model of three nanoobjects (NP-QD-NP) coupled by nonlinear dipole-dipole interactions in the presence of an external magnetic field. The nonlinear regime of this ensemble corresponds to the two-quantum processes of the QD biexciton decay in the case $|\delta| > \Omega_{1,2}, |\Delta|$, where $\delta = \varpi - \omega$ and $\Delta = \varpi - \omega_p$, ϖ is the spasing frequency, and $\Omega_{1,2}$ are the Rabi frequencies of dipole-dipole interactions between the QD and NPs. As a result of this nonlinear process, one can expect the appearance of strongly correlated plasmon pairs. In the technical framework, the presented spaser systems can be used to generate nonclassical states of the electromagnetic field at the nanoscale. Such systems can be integrated in the individual plasmonic waveguides [1] and plasmonic circuits for quantum information processing [2]. Nonlinear regimes of interaction between NP and QD can be realized, firstly, in the presence of a two-photon pump in the system and secondly, under the condition that the coupling energy between two electron-hole pairs is of the same order of magnitude as the internal coupling energy of a single pair. The energy of biexciton state XX differs from the double energy of the exciton X by the biexciton binding energy Δ_{2b} (see Fig. 1).

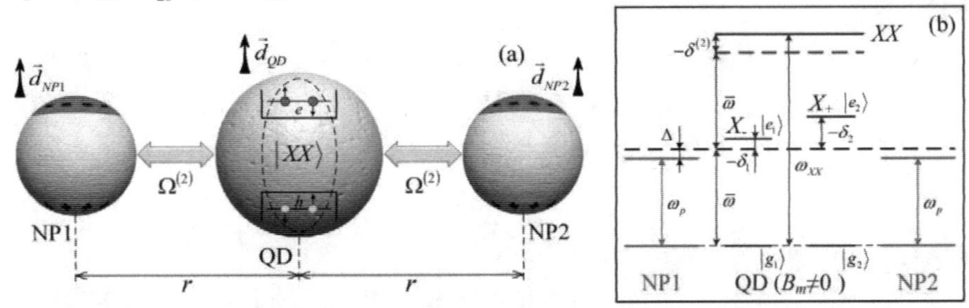

Fig. 1. (a) A model of a three-particle NP-QD-NP spaser. (b) The mapping of exciton and plasmon energy levels is in the presence of external magnetic field.

The Hamiltonian of exciton-plasmon interaction has the following form [3]:

$$H = \hbar\omega_{p1}\hat{c}_1^+\hat{c}_1 + \hbar\omega_{p2}\hat{c}_2^+\hat{c}_2 + \hbar\omega_{XX}D/2 + \hbar\Omega^{(2)}\left(\hat{c}_1\hat{c}_2\hat{S}^+ + \hat{c}_1^+\hat{c}_2^+\hat{S}\right), \quad (1)$$

where the last term in the brackets comprises the annihilation operator $\hat{S}$ of the biexciton XX state and the creation operators $\hat{c}_1^+$ and $\hat{c}_2^+$ of a pair of plasmons, whose energies differ slightly for the different intermediate levels X_+ (X_-) with frequencies ω_{X_+} (ω_{X_-}). The parameter $\Omega^{(2)} = \Omega_1\Omega_2/(2|\delta|)$ is the effective two-quantum Rabi frequency, D is the population imbalance. Parameters $\delta = \varpi - \omega_X$ and $\delta^{(2)} = 2\varpi - \omega_{XX}$ are the effective detunings, where $\omega_{XX} = \omega_{X_-} + \omega_{X_+} - \Delta_{2b}$ is the biexciton frequency. The Rabi frequency of interaction between the QD and NP is $\Omega_1 = \Omega_2 = \mu_{QD}/r^3\sqrt{\omega_p a_{NP}^3/(4\pi\hbar\varepsilon_0)}$, where a_{NP} is the NP radius, ω_p is the plasmon resonance frequency of NPs, μ_{QD} is the dipole moment. In this work we analyse the dynamics of the parameter

$$G_{12}^{(2)}(t,\tau) = \frac{\left\langle \hat{c}_1^+(t)\hat{c}_1(t)\hat{c}_2^+(t+\tau)\hat{c}_2(t+\tau)\right\rangle}{\left\langle \hat{c}_1^+(t)\hat{c}_1(t)\right\rangle\left\langle \hat{c}_2^+(t+\tau)\hat{c}_2(t+\tau)\right\rangle}, \quad (2)$$

which corresponds to the cross-correlation function and is a criterion for establishing correlations between plasmonic modes $\hat{c}_1$ and $\hat{c}_2$. In particular, the condition $G_{12}^{(2)} \equiv G_{12}^{(2)}(t,0) > 1$ is associated with the intermode plasmon bunching. Moreover, the violation of the Cauchy-Schwarz inequality $C \equiv \dfrac{G_{12}^{(2)}(t,0)^2}{g_1^{(2)}(t,0)\,g_2^{(2)}(t,0)} \le 1$ indicates the nonclassical character of the correlations, where $g_{1(2)}^{(2)} = G_{1(2)}^{(2)} + 1$. The maximum value of C parameter is $C_{\max} = 15.15$ and the stationary regime value is $C_{\rm st} = 1.2$, which demonstrates the nonclassical character of correlations between plasmon modes in spaser system. Thus, the main result of our simulation is a demonstration of the development of quantum correlations between two localized plasmonic modes and the possibility for formation of an entangled state of plasmons. Further development of this work may be aimed on a complex simulation of nonlinear collective processes [4] with non-classical states of surface plasmon-polaritons (SPP) [5] and SPP waves structures [6].

This work was supported by the Russian Foundation for Basic Research Grant No. 17-42-330029 and Moscow State Pedagogical University Postdoctoral Fellowships in Humanities 2018, Grant No. 0-1067.

References

1. C. Reinhardt, A.B. Evlyukhin, W. Cheng et al., J. Opt. Soc. Am. B **30**, 2898 (2013)
2. D.Yu. Fedyanin, D.I. Yakubovsky, R.V. Kirtaev, et al., Nano Lett. **16**, 362 (2016)
3. M. Yu. Gubin, A. V. Shesterikov, S. N. Karpov, A. V. Prokhorov, Phys. Rev. B **97**, 085431 (2018)
4. A.V. Golovanova, M.Yu. Gubin, M.G. Gladush, A.V. Prokhorov, Bull. Russ. Acad. Sci. Phys. **80**, 808 (2016)
5. A.B. Shesterikov, M.Yu. Gubin, M.G. Gladush, A.V. Prokhorov, JETP **124**, 18 (2017)
6. I.V. Dzedolik, V. Pereskokov, J. Phys.: Conf. Ser. **737**, 012006 (2016)

Time resolved femtosecond spectroscopy of nanocomposites

Oleg Khasanov[1,*], *Tatiana* Smirnova[2], *Olga* Fedotova[1], *Grigory* Rusetsky[1], *Vladimir* Gayvoronsky[3], and *Sergey* Pokutnyi[4]

[1]Scientific-Practical Material Research Centre NASB, 19 Brovki, Minsk 220072, Belarus
[2]International Sakharov Environmental University, 23 Dolgobrodskaya, Minsk 220070, Belarus
[3]Physics Institute NASU, 46 Nauky Ave., Kyiv 03680, Ukraine
[4]Chuiko Institute of Surface Chemistry, NASU, 17 Generala Naumova, Kyiv 03164, Ukraine

Abstract. Characterization methods of nanocomposites consisted of semiconductor metal-oxide quantum dots (QD) incorporated into a dielectric matrix have been elaborated on the base of time resolved four-wave mixing and photon echo. Large permanent dipole moment, inherent to QDs under study, local field effect, the QD spatial dispersion and distribution function of the transition dipole moments in QDs are taken into account. New responses at multiple frequencies in directions differed from spatial synchronism conditions of well-known signals have been predicted.

As it is well known, nanocomposites attract attention of scientists due to their unique properties. So, nanocrystals TiO_2 incorporated into potassium dihydrogen phosphate (KDP) crystals provide enhancement of the second harmonic generation efficiency at higher than 60% in comparison with pure KH_2PO_4. Macroscopic and microscopic features of such nanocomposite is actively discussed but the origin of observed effect is still under question. Bearing in mind high potential for frequency up- and down-conversion of nanocomposites composed of semiconductor metal-oxide quantum dots (QD) incorporated into a dielectric matrix we study exciton states in these materials and elaborate characterization methods accounting for large permanent dipole moment (PDM) of quantum dots (QD).

We dwell on the nanocomposite consisted of semiconductor spherical QD with the permittivity ε_2 surrounding insulating matrix with the permittivity ε_1, and suppose, that $\varepsilon_2 >> \varepsilon_1$.In such a system QD-matrix interface forms a potential barrier to electrons not allowing them to penetrate into the QD. The Coulomb interaction of the electron with the QD-matrix interface leads to localization of the electron in the potential well near the QD surface. With increasing QD radius *a,* so that $a >> a_b$, where a_b is the exciton Bohr radius, the spherical QD-matrix interface transforms into a planar interface between the semiconductor and the matrix. The exciton formed from the spatially separated electron and hole becomes two-dimensional (2D) with maximal binding energy. Another dielectric

* Corresponding author: olkhas@mail.ru

enhancement effect is originated from the field produced by the nanoparticles in matrix, when $\varepsilon_2 >> \varepsilon_1$ Under these conditions, rich structure of exciton levels is formed with a multitude of possible transitions. It should be noted that in such a nanosystem the PDM and transition dipole moment can exceed the same parameters for volume samples by two orders of magnitude. As a sequence, methods of time resolved laser spectroscopy should be modified accounting PDM impact. Thus, our task reduces to the elaboration of characterization technique of noncentrosymmetric multilevel system by femtosecond pulses. Below we more specifically consider features of time resolved four wave mixing (TRFWM) and photon echo (PE) methods for nanoparticles embedded in noncentrosymmetric matrices with quadratic nonlinearity.

For simplicity, we consider three-level system, which is in resonance with pump pulses at the lowest transition between exciton states. Due to the PDM interaction with the light pulses the Hamiltonian of the light-matter interaction contains fast oscillating diagonal elements that makes impossible traditional theoretical technique application to calculate responses of medium on pulse excitation. After unitary transformation

$$\hat{L} = \begin{pmatrix} \exp(i\lambda) & 0 & 0 \\ 0 & \exp(-i\lambda) & 0 \\ 0 & 0 & 1 \end{pmatrix}$$

where $\lambda = \frac{p_{11}}{\hbar} \int E_i dt = \sum_{m=1}^{\infty} \frac{p_{11} E_i}{\hbar\omega} \sin\varphi(z,t), \varphi(z,t) = (\omega t - \vec{k}_i \vec{r})$, we arrive at Hamiltonian describing QD interaction with a number of fields at multiple frequencies. Here p_{11} is the PDM, E_i, ω and $\dot{k}_i$ are the envelope, carrier frequency and wave vector of *i*-th pulse, correspondingly. Applying successively a number of unitary transformations, we obtain Hamiltonian for generalized two-level system in rotating reference framework. This Hamiltonian does not contain fast oscillating terms and describes one- and two-quantum transitions. While one-quantum transitions occur in the field of excitation pulse E_i two-quantum transitions are induced by its components $E_i J_s(a)$ and $E_i J_{s-1}(a)$ at multiple frequencies $(s+1)\omega$ and $s\omega$ due to the PDM contribution. Elaborated approach allows us to deduce the Bloch equations for the QD with PDM in slowly varying envelope approximation and to solve them by means of well-known computational procedures. Besides resonant component macroscopic photoinduced polarization contains multiple harmonics, which can be sources of responses at corresponding frequencies in the nanocomposite. We have calculated TRFWM and PE signals under noncollinear excitation taking into account excitation induced shift of absorption frequency, the QD spatial dispersion and distribution function of the transition dipole moments in QDs. In addition to signals at resonant frequency under well-known spatial synchronism conditions, we predict observation of responses on two-pulse excitation of nanocomposite at multiple frequencies. The observation time and spatial synchronism conditions of new responses will differ from signal in $(2\dot{k}_2 - \dot{k}_1)$ direction. The signals at multiple frequencies and terahertz responses are originated from the PDM interaction with excitation pulses. The most intensive signals among all responses will be at double and terahertz frequencies. Signals related to local field effect are discussed. It should be noted that for short time delay between pulses the nonresonant nonlinear matrix will contribute to the TRFWM signals mainly to resonant, second and terahertz harmonics.

Emerging upconversion nanoparticles for industry and biomedical application

Evgeny Khaydukov[1,*], *Vasilina* Rocheva[1], *Alexander* Savelyev[1], *Kirill* Khaydukov[1], *Ilya* Asharchuk[1], *Andrey* Nechaev[1,2], *Dmitry* Khochenkov[1,3], *Anastasia* Sochilina[1,4], *Vladimir* Semchishen[1] and *Alla* Generalova[1,4]

[1]Federal Scientific Research Centre "Crystallography and Photonics" of the Russian Academy of Sciences, Leninsky pr. 59, 119333 Moscow, Russia
[2]Institute of Fine Chemical Technologies, Moscow Technological University, 86 Prospect Vernadskogo, 119571 Moscow, Russia
[3]N.N. Blokhin Russian Cancer Research Center, 115478, Moscow, Russia
[4]Shemyakin-Ovchinnikov Institute of Bioorganic Chemistry of the Russian Academy of Sciences, 16/10 Miklukho-Maklaya, 117997 Moscow, Russia

Abstract. In recent years, the overwhelming majority of the upconversion nanoparticles (UCNPs) prominent applications have originated from their unique luminescent properties. Due to original properties of inorganic UCNPs they attract the interest in numerous fields. We discussed a number of UCNP assisted techniques, such as biomedical imaging, therapy agents, anti-counterfeit labels and 3D printing, showing highly versatile and translatable UCNP photoluminescent nanotechnology for the applications in industry and biomedicine.

The main engineering advantage of UCNP platform, unlike conventional luminescent labels, is based on the nature of the anti-Stokes luminescence. UCNPs consist of an inorganic host matrix (usually $NaYF_4$, YF_3, $LiYF_4$) co-doped with lanthanide ions act as the sensitizer (e.g., Yb^{3+} or Nd^{3+}) and as the activator (e.g., Er^{3+}/ Tm^{3+} or Ho^{3+}). The network of closely spaced sensitizer-ions resonantly absorbs NIR radiation and non-radiatively couples excitation energy to neighboring activator -ions. Activators provide the multiple photon absorption due to their ladder-like arranged levels and emit upconverted photons [1].

1 Near-infrared polymerization and security printing

We demonstrated the effective and straightforward strategy for the NIR-activated polymerization of photocurable compositions containing UV-emitting upconversion nanomaterials. For this aim, the rationally-designed core/shell upconversion nanoparticles $NaYF_4:Yb^{3+},Tm^{3+}/NaYF_4$, with the distinct ultraviolet-emitting lines have been used. Their ability to activate commercially available photoinitiators in the process of radical polymerization of light-sensitive resins at moderate NIR light intensities and irradiation

* Corresponding author: khaydukov@mail.ru

doses opens unique possibilities for the implementation of novel NIR triggered photopolymerization technique. This technique allows fabrication of 3D polymeric structures inside the volume of photocurable compositions.

We developed invisible biocompatible inks visualized under infrared irradiation for security printing applications. To produce stable water inks the surface of UCNPs was modified by amphiphilic polymer. Undetectable under daylight security printing was performed by ink-jet printing on paper. Upconversion of the printed patterns under near infrared laser irradiation was observed and imaged. Higher levels of security were achieved based on optical multiplexing capable combinations of Ln^{3+} in nanoparticles. The developed technology is promising for fast, easy and low cost inkjet printing for anti-counterfeit and security applications.

Fig. 1. Image of 3D polymer microstructure obtained by NIR-light-activated photopolymerization (left). The graphic image printed with upconversion inks (middle). UCNP biovisualization - the tumor exhibited high luminescent contrast (red color) demonstrating superiority of the UCNP-assisted imaging (right).

2 Bioimaging and Photodynamic therapy

Photodynamic therapy (PDT) is a clinical tumor treatment that utilizes cytotoxic reactive oxygen species (ROS), generated by photosensitizer, which is excited by external light source. However, conventional PDT is limited by the penetration depth of light required for photosensitizer activation. We demonstrate a solution to this problem by the rational photoluminescent nanotechnology design. Specifically, upconversion nanoparticles (UCNPs) offer a lucrative possibility to kindle UV or visible light at the sub-centimetre depth in tissue, owing to their property to convert infrared radiation at the wavelength in the biological tissue transparency window 975 nm. We designed upconversion nanoparticles (UCNP) of the composition $NaYF_4{:}Yb^{3+}{:}Tm^{3+}$ that yielded UV/blue emission bands critical for the photosensitising of vitamin B_2. We achieved a high Förster resonance energy transfer (FRET) efficiency from UCNPs to vitamin B_2, which facilitates a large production of cytotoxic singlet oxygen and thus an enhanced PDT efficacy. Design UCNPs was used as PL probes for tumor labelling. The comparison of the image acquisition in the Stokes and anti-Stokes channels shows much greater contrast in the anti-Stokes channel. This clearly demonstrates the advantage of UCNP-assisted imaging.

This scientific work has been partly supported by Grants RSF No. 16-13-10528 (in the part of UCNP assisted NIR-activated polymerization), RFBR No. 17-00-00122 (K) (in the part of UCNP assisted bioimaging and therapy).

References

1. A. N. Generalova, B. N. Chichkov, E. V. Khaydukov, Adv. Colloid Interface Sci. **245**, 1 (2017).

New schemes of compact modal LC modulators for control of light spatial structure

Svetlana Kotova*, *Alexandra* Mayorova, and *Sergey* Samagin

Lebedev Physical Institute RAS, Samara branch, 221, Novo-Sadovaya Str., Samara, 443011 Russia

Abstract. The results of numerical simulation of new schemes of compact modal LC spatial modulators are presented. Special configurations of contact electrodes and predetermined values of applied voltages provide the jump in the profile of phase delay of light passing through the modulator or reflecting from it. This leads to the appearance of a vortex field. Ring-shaped light fields and light fields with intensity distribution in the form of two maxima, which rotate in the transverse plane during propagation were obtained. These devices may be valuable for use in the spectroscopy of single molecules.

Structured light has a huge number of applications, including microscopy and spectroscopy [1]. Formation of light fields with various (arbitrary) intensity distributions and dynamic control of their parameters is possible using multi-pixel LC SLM. For practical applications the reducing cost of modulators as well as the simplification and reducing the sizes of the optical schemes on their base are of interest. In this paper we present new schemes of compact modal LC light modulators which can form certain types of vortex light fields.

The modal principle of control has been proposed in studies [2, 3]. The important feature of those modulators is the presence of additional homogeneous transparent high-resistance layer. This layer covers the low-resistance coating with the hole and performs the function of a control electrode. Tunable spherical and cylindrical lenses were designed in this way. Next, we developed LC focusator (or 4-channel LC modulator) [4, 5]. By using LC focusator as an optical transparent it is possible to focus light into a point spot or a line segment and to form transverse distributions of intensity. Shaped beams can be of various forms, with sizes, position and orientation controlled by means of applied voltages [6 - 8]. Since only a few control electrodes and low control voltages are used in modal LC devices they are easy to control, have relatively low cost and are easy to manufacture.

To obtain light fields with angular momentum, it is necessary to provide a jump in the phase profile of the light passing through the device. This can be done with the special geometry of electrodes. It is found that axially symmetric light intensity distributions can be obtained for two electrode configurations. In the first case, the contact electrode on one of substrates is divided in half, and in the second one, two narrow stripe electrodes are applied on one of the substrates. Thus, a spiral focusator and spiral phase plate can be realized. Selecting the amplitudes of the potentials in such a way that the magnitude of the jump in the

* Corresponding author: kotova@fian.smr.ru

phase profile varies by a factor of 2π, it is possible to control the diameter of the ring (Fig. 1). It is also possible to move the light ring in the observation plane. This case may be of interest for example for STED-nanolithography.

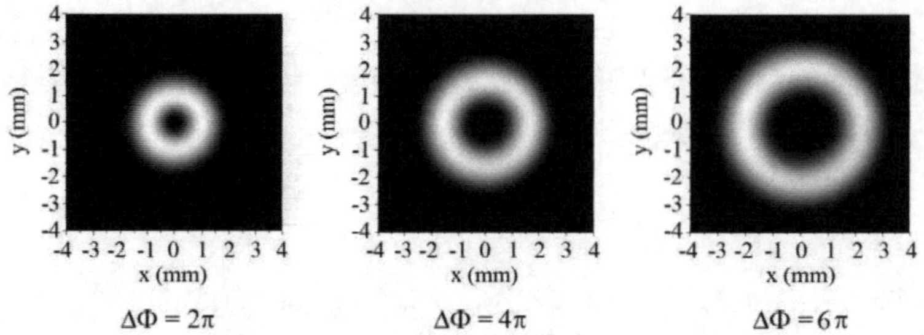

Fig. 1. Dynamic control of the formed fields. Distributions of the intensity in the observation plane for different potentials producing the phase jump $\Delta\Phi$.

If the conductive coating and contact strip electrodes on one of the substrates are divided in the middle by a narrow nonconducting transparent strip, then on this substrate four contact electrodes are formed. With the help of such a device, one can obtain an intensity distribution with two maxima that rotate during propagation. Incident light of different curvatures will produce maxima at different positions in the set observation plane (Fig. 2). This configuration of the modulator can be used in fluorescent nanoscopy [9].

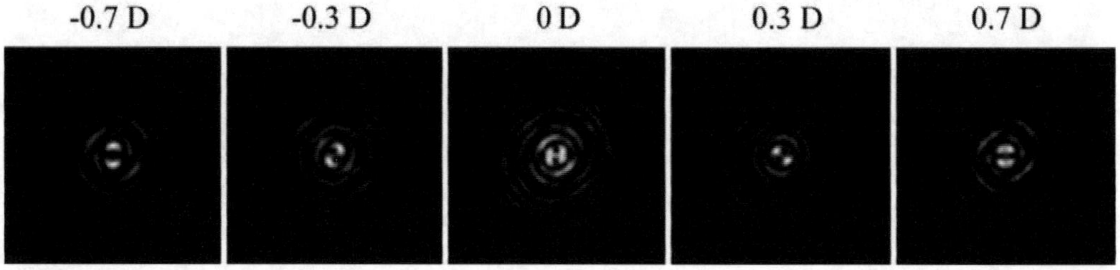

Fig. 2. Intensity distribution of the light in the observation plane for different curvatures (in dioptres) of the incident wave front.

This research is financially supported by the RFBR (projects Nos. 16-02-01055, 16-29-11809 and No. 16-42-630773).

References

1. H. Rubinsztein-Dunlop et al., Journal of Optics, **19**, 013001 (2017)
2. A.F. Naumov, Proc. P.N. Lebedev Physical Institute **217**, 182 (1993)
3. A.F. Naumov, M.Y. Loktev, I.R. Guralnik and G, Vdovin, Opt. Lett., **23**, 992 (1998)
4. S.P. Kotova, V.V. Patlan and S.A. Samagin, Quantum Electronics, **41**, 58 (2011)
5. S.P. Kotova, V.V. Patlan and S.A. Samagin, Quantum Electronics, **41**, 65 (2011)
6. A.V. Korobtsov, S.P. Kotova, N.N. Losevsky, A.M. Mayorova and S.A. Samagin, Quantum Electronics, **44**, 1157 (2014)
7. S.P. Kotova, A.M. Mayorova and S.A. Samagin, Journal of Optics **17**, 055602 (2015)
8. S.P. Kotova, A.M. Mayorova and S.A. Samagin, Journal of Opt. Tech., **84**, 323 (2017)
9. A.V. Naumov, Phys. Usp., **56**, 605 (2013)

Experimental realization of revival of silenced echo memory protocol in optical cavity

Mansur Minnegaliev*, *Konstantin* Gerasimov, *Ravil* Urmancheev, and *Sergey* Moiseev

Kazan Quantum Center, KNRTU-KAI, 420113, K. Marx Str. 10, Kazan, Russia

Abstract. We demonstrated a photon echo quantum memory for weak input optical pulses on the ROSE protocol in a Tm^{3+}:$Y_3Al_5O_{12}$ crystal placed in impedance-matched optical cavity. The quantum efficiency of 21% for a storage of time of 36 μs was achieved for single light pulses.

Active development of optical quantum technologies including optical quantum computing and long range quantum communications stimulates the creation of quantum memory (QM). The creation of highly-efficient QM will not only significantly expand the capabilities of these technologies, but also leads to fundamental impact on the creation of new directions in their development [1]. In last decade, there was proposed and experimentally realized a number of protocols of QM [2]. The schemes based on photon echo in solid state systems [3] demonstrated very promising results for achieving high quantum efficiencies and information capacity. Herein, the revival of silenced echo (ROSE) memory protocol seems especially attractive due to the possible use of natural inhomogeneously broadened line [4-6] that could significantly simplify practical implementation of QM. It was also recognized that the photon echo approach in free space schemes needs for a large optical density and requires backward spatial geometry for efficient quantum retrieval. However, these limitations can be eliminated by placing a QM cell in an optimal optical cavity [7,8]. Crystals doped by rare-earth ions are very attractive for this scheme due to its parameters of homogeneous and inhomogeneous broadening of the optical transitions [9]. By placing this crystal into impedance matched optical cavity we can increase the light-matter interaction up to complete absorption of input optical signals which leads to higher efficiencies [7,8]. In recent years there were several experimental demonstrations of photon echo QM in optical cavities [10,11,12].

In this work for the first time we experimentally realized ROSE protocol in a Tm^{3+}:$Y_3Al_5O_{12}$ crystal placed in a concentric optical cavity. As in previous work [6] we used $^3H_6$-$^3H_4$ optical transition (λ = 793 nm) of thulium ions with concentration of 0.1% and orthogonal geometry for propagation of signal and control fields in the cavity. The concentric optical cavity was constructed from two spherical dielectric mirrors with reflection coefficients of R_1=70% for the front (coupling) mirror and $R_2\approx$100% for back mirror. The overall cavity length was ~90 mm which corresponded to 1.65 GHz free-spectral range. The impedance-matching condition for the ideal cavity with coupling mirror reflection R_1 and without any losses inside is achieved when weak total absorption α of the crystal with length L satisfies $2\alpha L=-\ln(R_1)$. In our case this condition is satisfied for $\alpha L\approx0.15$. So the experiment was performed on absorption line wing because αL is close to

* Corresponding author: minnegaliev.mansur@yandex.ru

1.4 at the line center. The crystal was placed in Montana Instr. closed-cycled cryostation and cooled down to the temperature T = 4 K. Typical behavior of the input light pulse and detected revived silenced echo is depicted in Fig. 1. One can see an absence of total impedance-matching condition because 7.5 % percent of the input pulse was reflected. This could be caused by misalignment of the cavity. For no cavity case, the unabsorbed (reflected) signal could be 85% in this scheme. So the impedance-matched cavity provides 10-fold enhancement of the signal absorption. Using this property, we obtained retrieval efficiency of 21% of input signal pulse for a storage time of 36 μs which is the best result for ROSE experiments and third in implementation of QM in optical cavity.

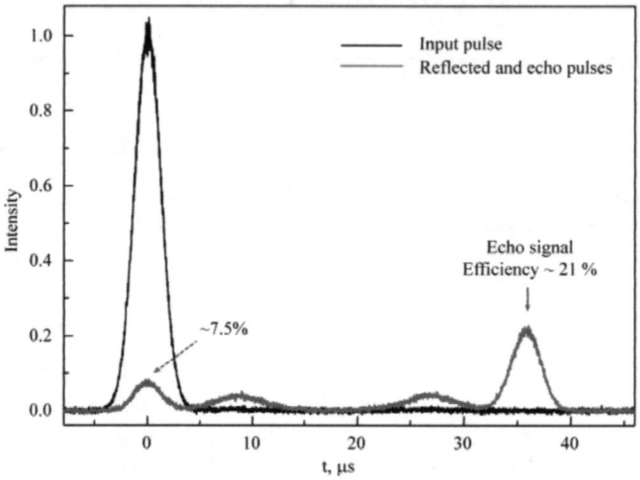

Fig 1. The signal of the revived silenced echo obtained by using Tm^{3+}:$Y_3Al_5O_{12}$ crystal in impedance matched optical cavity. The signals from scattered π-pulses (red curve) can be seen between the input pulse and echo signal. The echo signal efficiency was 21% with storage time 36 μs.

This work was supported by Russian Scientific Fund, Grant No. 14-12-01333 P.

References

1. F. Bussières, N. Sangouard, M. Afzelius, H. de Riedmatten, C. Simon, and W. Tittel, J. Mod. Opt., **60**, 1519 (2013)
2. K. Heshami et al., J. Mod. Opt., **63**, 2005 (2016)
3. W. Tittel et al., Las. Photon. Rev., **4**, 244 (2009)
4. V. Damon, M. Bonarota, A. Louchet-Chauvet, T. Chanelière, and J.-L. Le Gouët, New J. Phys., **13**, 93031 (2011)
5. S.A. Moiseev, Phys. Rev. A, **83**, 012307 (2011)
6. K.I. Gerasimov, M.M. Minnegaliev, S.A. Moiseev, R.V. Urmancheev, T. Chanelière, and A. Louchet-Chauvet, Opt. Spectrosc., **123**, 211 (2017)
7. S.A. Moiseev, S.N. Andrianov, and F.F. Gubaidullin, Phys. Rev. A, **82**, 22311 (2010)
8. M. Afzelius and C. Simon, Phys. Rev. A, **82**, 022310 (2010)
9. C.W. Thiel, T. Böttger, and R.L. Cone, J. Lumin., **131**, 353 (2011)
10. M. Sabooni, Q. Li, S. Kröll, and L. Rippe, Phys. Rev. Lett., **110**, 133604 (2013)
11. P. Jobez, I. Usmani, N. Timoney, C. Laplane, N. Gisin, M. Afzelius, New J. Phys. **16**, 083005 (2014)
12. R.A. Akhmedzhanov, L.A. Gushchin, A.A. Kalachev, N.A. Nizov, V.A. Nizov, D.A. Sobgayda, and I.V. Zelensky, Las. Phys. Lett., **13**, 115203 (2016)

Live-cell nanoscopy enabled with transient labeling and the control of fluorophore blinking

Alexander Mishin*, *Maxim* Perfilov, *Alexey* Gavrikov, *Anastasia* Mamontova, *Alexey* Bogdanov, and *Konstantin* Lukyanov

Shemyakin-Ovchinnikov Institute of Bioorganic Chemistry of the Russian Academy of Sciences 16/10 Miklukho-Maklaya, Moscow, 117997, Russian Federation

Abstract. Live-cell super-resolution of proteins labeled with genetically encoded fluorescent tags is a challenging task because of the imperfect labeling and the inevitable deterioration of the signal in the course of the experiment. Incomplete maturation of the covalently attached fluorescent tags, inefficient photoconversion, and photobleaching further complicate prolonged live-cell nanoscopy. We have implemented two strategies for lowering the photodamage: ensuring the dynamic replacement of damaged molecules and establishing conditions for the robust intrinsic blinking of the tags at lower illumination powers.

Typical super-resolution localization microscopy setup requires illumination powers at the levels harmful to living cells [1]. These harsh conditions are necessary for both squeezing out the maximum possible number of photons from the fluorophore and for the robust blinking of the dye. Alternative means of achieving blinking established for fixed samples, such as the use of strong reductants, are also inadequate for live-cell experiments. Therefore, the photoactivatable and photoswitchable fluorescent proteins (such as Dendra2, mEos3.2, Dronpa, PA-GFP) currently occupy the niche of fluorescent tags for live-cell super-resolution. For that type of probes, the precise control of the number of fluorophores in the light-emitting state is achievable via the additional light source, typically in the near-UV range. We have recently shown that the intrinsic blinking of some genetically encoded proteins can be robust enough even at low-light setup to allow nanoscopy applications [2]. Still, the labeling quality inevitably deteriorates in the course of the experiment due to the depletion of the pool of the available fluorophores. Unfinished maturation of the covalently attached fluorescent marker, incomplete photoconversion, and photobleaching hinder prolonged live-cell nanoscopy and reduce effective labeling density (Fig. 1A).

The general principles of PAINT (Point Accumulation for Imaging in Nanoscale Topography) resolve many challenges in super-resolution microscopy by replenishing the photobleached or defunct dye from the pool of undamaged molecules [3]. We have recently shown the protein-PAINT with a specific transient binding of cell-permeable fluorogenic dyes to genetically encoded protein tags, engineered from bacterial lipocalins [4]. In protein-PAINT, the concentration of externally added chemical dye controls the number of blinking events (Fig. 1B).

* Corresponding author: mishin@ibch.ru

While the photon budget of these dyes does not exceed one of the fluorescent proteins, an almost infinite supply of undamaged dyes allows for nanoscopy even at the low-light regime.

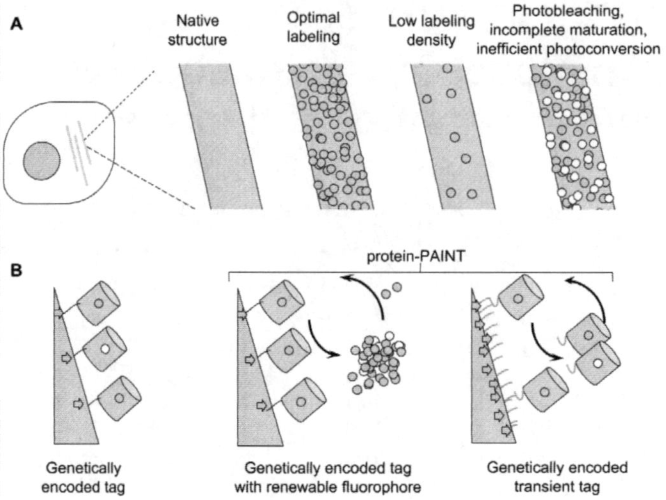

Fig. 1. Challenges of live-cell super-resolution with genetically encoded tags. (A) scheme depicts an enlarged region of the cytoskeleton, the problems in labeling are stated; (B) scheme depicts an enlarged region of a cellular structure, labeled with (left-to-right) a fluorescent protein, the protein-PAINT two-component system with a pool of renewable fluorophores, or with entirely genetically encoded protein-PAINT implementation based on transient protein-protein interactions.

Here we further extend the range of compatible fluorogenic dyes and test other applications of the labeling system, including FRET. We then show the Protein-PAINT implementation based on fluorescent protein and the transient heterodimerization of artificial peptides. Similarly to the protein-PAINT with synthetic dyes, we confirm the improvement in the signal photostability and the labeling density in confocal and super-resolution regimes. Also, we provide the data on the effect of the amino acid residues near the chromophore on blinking of photoactivatable green fluorescent proteins. In conclusion, we show how all the techniques mentioned above could be combined to allow prolonged low-light nanoscopy in living cells with minimal photodamage.

This work was supported by the Russian Science Foundation grant 16-14-10364.

References

1. S. Wäldchen, J. Lehmann, T. Klein, S. van de Linde, M. Sauer, Sci. Rep. **5**, 15348 (2015)
2. N. V. Klementieva, A. I. Pavlikov, A. A. Moiseev, N. G. Bozhanova, N. M. Mishina, S. A. Lukyanov, E. V. Zagaynova, K. A. Lukyanov, A. S. Mishin, Chem. Commun., **53**, 949 (2017)
3. A. Sharonov, R. M. Hochstrasser. Proc. Natl. Acad. Sci. U.S.A, **103**, 18911 (2006)
4. N. G. Bozhanova, M. S. Baranov, N. V. Klementieva, K. S. Sarkisyan, A. S. Gavrikov, I. V. Yampolsky, E. V. Zagaynova, S. A. Lukyanov, K. A. Lukyanov, A. S. Mishin, Chem. Sci, **8**, 7138 (2017)

Gel electrophoresis and hybridization of DNA-functionalized quantum-dots as part of a study of linear quantum-dot chains

Masaru Oda[1,2,*], *Kosuke* Sasano[2] , *Akira* Nishi[2] , *Tamotsu* Zako[3] , and *Toshiro* Tani[4]

[1]Department of Basic Sciences, Faculty of Engineering, Kyushu Institute of Technology, 1-1 Sensui-cho, Tobata-ku, Kitakyushu, Fukuoka, 804-8550 Japan
[2]Department of Applied Science for Integrated System Engineering, Graduate School of Engineering, Kyushu Institute of Technology, 1-1 Sensui-cho, Tobata-ku, Kitakyushu, Fukuoka, 804-8550 Japan
[3]Department of Chemistry and Biology, Graduate School of Science and Engineering, Ehime University, 2-5, Bunkyo-cho, Matsuyama, Ehime, 790-8577 Japan
[4]Department of Applied Physics, Tokyo University of Agriculture and Technology, Naka-cho 2-24-16, Kogane-i, Tokyo, 790-8577 Japan

Abstract. We report the results of our investigation of agarose gel electrophoresis and hybridization of mercaptopropionic-acid-capped quantum dots (MPA-capped QDs) conjugated with short single-stranded DNA, as part of a study of closely spaced linear QD chains. The electrophoresis results suggest that migration speed and separation resolution depend on the amount of MPA on the QD surface. The QDs separated by electrophoresis according to DNA valence, i.e., the amount of DNA per particle, were used as building blocks for the synthesis of linear QD oligomers.

1 Introduction

Closely spaced quantum dot (QD) oligomers, such as dimers and trimers, have attracted attention over the last two decades due to their optical properties, which are influenced by energy transfer and/or electronic coupling among constituent QDs [1-3]. However, QD oligomer optical properties are still not thoroughly understood, in part because it is difficult to fabricate well-defined QD-oligomer structures, especially linear chains composed of three or more QDs. We recently conducted an investigation to improve QD oligomer synthesis using a DNA hybridization technique. Here, we present the results of this investigation.

2 Experimental

QD oligomers were fabricated via the following steps. First, CdSe/ZnS core-shell QDs capped with trioctylphosphineoxide were synthesized via pyrolytic decomposition of

* Corresponding author: odamasa@mns.kyutech.ac.jp

organometallic compounds [4]. Next, the QDs were rendered water-soluble by ligand exchange with mercaptopropionic acid [5]. The QDs were conjugated with thiolated short single-stranded (ss) DNA (15 mer) in a borate buffer solution (pH 8.3). The DNA-functionalized QDs were separated using agarose gel electrophoresis according to DNA valence (N_{DNA}), i.e., the amount of DNA per particle. The QDs for the formation of linear QD-oligomers (N_{DNA} = 1 or 2, as shown in Fig. 1(a)) were extracted from the gel and linked with similarly prepared QDs conjugated with complementary DNA, as illustrated in Fig. 1(b). The structures of the oligomers were characterized by transmission electron microscopy (TEM). Absorption and photoluminescence (PL) spectra were measured.

3 Results and Discussion

Figure 1(c) shows a typical PL image of the agarose gel containing the QDs conjugated with ssDNA after electrophoresis. The QDs were separated into several bands. We confirmed that the migration speed decreases as the DNA valence increases, as shown in Fig. 1 (a) [6]. On the other hand, the migration order is reversed and the bandwidth, i.e., separation resolution, is improved as the amount of MPA on the QD surface decreases. Details of the separation by electrophoresis will be described in the presentation.

Figure 1(d) shows typical TEM images of QD-oligomers obtained after hybridization. Even though precise control of the shape and particle-number was not achieved, it was confirmed that closely spaced QD-oligomers could be obtained by the hybridization.

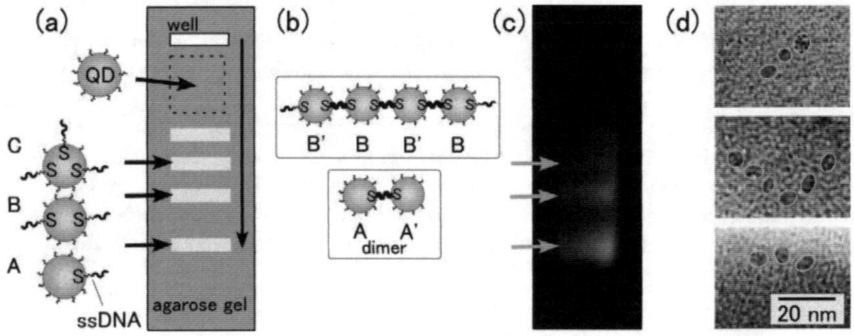

Fig. 1. Schematic diagram of (a) the agarose gel electrophoresis of the DNA-functionalized quantum dots (QDs) (b) QD chains linked by double-stranded DNA. (c) Photoluminescence (PL) image of the agarose gel (4% NuSieve GTG agarose; Lonza) under 365-nm light irradiation. The molar ratio of the QD to DNA was 1:8. (d) Transmission electron microscopy (TEM) images of QD oligomers. The lines represent the particle outlines.

References

1. R. Koole, P. Liljeroth, C. Donega, D. Vanmaekelbergh, and A. Meijerink, JACS, **128**, 10436 (2006)
2. B. K. Hughes, J. L. Blackburn, D. Kroupa, A. Shabaev, S. Erwin, A. Efros, A. Nozik, J. Luther, and M. Beard, JACS **136**, 4670 (2014)
3. G. Tikhomirov, S. Hoogland, P. Lee, A. Fischer, E. Sargent, and S. Kelley, Nature Nanotechnology, **6**, 485 (2011)
4. K. Hashizume, M. Matsubayashi, M. Vacha, and T. Tani, J. Lumin., **98**, 49 (2002)
5. B. Pong, B. Trout, and J. Lee, Langmuir, **24**, 5270 (2008)
6. M. Oda, K. Sasano, A. Nishi, T. Zako, and T. Tani, submitted

Thin layer fluorescence microscopy based on one-dimensional photonic crystal

Kirill Prusakov[1,2,*], *Dmitry* Basmanov[1], and *Dmitry* Klinov[1]

[1]Federal State Budgetary Institution Federal Research and Clinical Center Of Physical-Chemical Medicine, Federal Medical Biological Agency, Laboratory of Medical Nanotechnologies, 119435 Malaya Pirogovskaya 1a, Moscow, Russia
[2]Moscow institute of physics and technology, Departure of Biological and Medical Physics, 141701 Institutskiy per. 9, Dolgoprudny, Moscow Region, Russia

Abstract. A new method of specimen illumination for wide-field fluorescence microscopy has been presented. This method allows to excite the fluorescence in a thin near-surface layer of the studied object. As a result, the captured images have greater contrast and signal-to-background ratio in comparison with the epifluorescence ones. The long-range surface waves in one-dimensional photonic crystal have been used to localize the electromagnetic field exciting the fluorescence. An experimental setup has been created to excite the surface waves and obtain images of the objects from the near-surface layer. For an illustration of the possibilities of our method, we conducted several experiments with specimens that are typical for fluorescence microscopy, such as bacteria and eukaryotic cells.

1 Introduction

One-dimensional photonic crystal (1D PC) is a multilayer meta-structure with periodically changing refractive indices of the layers. As shown by V. Konopsky et al. [1], it is possible to create the special conditions for existence of one or several modes of long-range surface optical range electromagnetic waves on the surface of 1D PC. The main idea of our work [2] is to use a small penetration depth of the long-range surface waves to improve the contrast and signal-to-background ratio of the images that could be captured with the help of a wide-field fluorescence microscope. One of the most frequently used methods of such kind image improvement is the so-called Total Internal Reflection Fluorescence Microscope (TIRFM). However usually, this approach requires permanent fine-tuning of optical setup and works only with high numeric aperture lenses. It will be shown that our method does not have such limitations.

2 Materials and Methods

The 1D PC structure was calculated using the method of complex impedance [3]. The dispersion characteristic of the calculated structure corresponded to the excitation spectra of

* Corresponding author: kaprusakov@gmail.com

green fluorescent protein (GFP). This structure provides an opportunity to initiate long-range surface waves with 473 nm wavelength on the 1D PC surface. The PC was fabricated using the magnetron sputtering and consisted of 3 layers of SiO_2 and Ta_2O_5 (134.4 nm and 78.4 nm thickness respectively) and a final layer of SiO_2 300 nm thick. In order to excite the long-range surface waves, an experimental setup based on the Kretschmann configuration with a cylindrical lens was created (see Fig. 1). This configuration does not require permanent fine-tuning of the laser illumination angle.

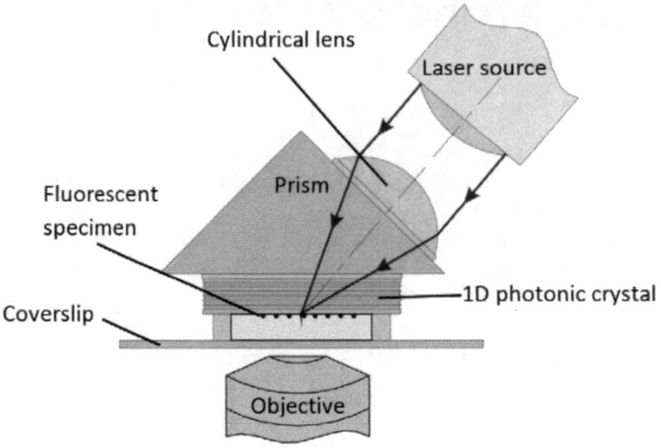

Fig. 1. An experimental setup based on the Kretschmann configuration.

Laboratory mutant strain of GFP labeled *E. coli* and a mutant line of GFP labeled HeLa cells were used as the representative specimens for fluorescent microscopy. Cells were fixed on the PC surface. The images of the PC surface were captured with the help of a wide-field fluorescence microscope.

3 Results and Conclusion

The experiments showed that fluorescence is excited only inside a thin near-surface layer (~150 nm) of a specimen. Moreover, in the obtained images the signal-to-background ratio is on the average 7 times greater than the value of this parameter in the epifluorescent images of the same specimens. Thus, the localization of the initiating radiation reduces a background signal from the specimen bulk. As a result, it became possible to study a more detailed structure of a specimen. This new method of specimen illumination can be used as a cheap and simple alternative to the TIRFM method. Also, it potentially can be combined with a surface waves biosensor [4].

This work was supported by the Russian science foundation (project № 17-75-30064).

References

1. V.N. Konopsky, E.V. Alieva, Phys. Rev. Lett., **97**, 253904 (2006)
2. K.A. Prusakov , D.V. Basmanov, D.V. Klinov, Patent RU 2626269 (2017)
3. V.N. Konopsky, New J. Phys., **12**, 093006 (2010)
4. M. Prokofjeva, V. Tsvetkov, D. Basmanov, A. Varizhuk, M. Lagarkova, I. Smirnov, K. Prusakov, D. Klinov, V. Prassolov, G. Pozmogova, S. N. Mikhailov, Nucleic Acid Ther., **27**, 56 (2017)

Plasmonic magneto-optical nested 2D nanostructures: tailoring responses through effective refractive index

Alexey Shaimanov[1,2,*] and *Alexander* Baryshev[1,3]

[1]All-Russia Research Institute of Automatics, Moscow 127055, Russia
[2]Moscow State University, Moscow 119991, Russia
[3]Moscow Institute of Physics and Technology, Dolgoprudny 141700, Russia

Abstract. Spectra of a nested square two-dimensional lattice of metal nanospheres encased in a magneto-optical host were studied. We show that the magneto-optical response of the nanostructures considerably increases due to the plasmon resonances. Moreover, the optical and magneto-optical responses can be strongly altered with a negligible change in structural parameters.

Nanophotonics encompasses studies on artificial nanostructures that support various optical resonances [1]: surface propagating plasmon resonance (SPR) and localized surface plasmon resonances (LSPR). These plasmonic resonances in various magneto-optical (MO) plasmonic systems are useful for the control of light intensity or polarization. The enhancement of polarization rotation is observed in the polarization rotation spectra of plasmonic MO nanostructures and should be understood as a reciprocal phenomenon [2], however, can be useful for a tunable waveplate [3].

To demonstrate a new view on plasmonic MO nanostructures, we introduce a nested square two-dimensional (2D) lattice of gold (Au) nanospheres encased in a magnetic host (a layer bismuth-substituted yttrium iron garnet, Bi:YIG). We analyze spectra of 2D Au-Bi:YIG nanostructures with square lattices having different periods and discuss LSPRs on single nanospheres and their ensemble. Then, we select a 2D lattice and insert additional nanospheres with radius **R**, composing another square lattice, into it, thus forming a nested 2D lattice. We will show that the number and spectral positions of the LSPR bands in spectra of the nested Au-Bi:YIG nanostructure are governed by the radius of nanospheres, the MO response can be enhanced at frequencies of LSPRs and it becomes extremely sensitive to the radius of the nanospheres. Two main features are in spectra of such strustures: LSPR (**D**) band is defined by the lattice period **D** and has a large spectral shift as **D** changes and LSPR on single spherical nanoparticle in the short-wavelength (SW) band at approximately 600 nm (see Fig. 1(a)).

We compare MO responces of the Au–Bi:YIG nanostructures with square and nested lattices and find interesting features. Unexpectedly, the sign of the angle of polarization

* Corresponding author: shaymanov.aleksey@physics.msu.ru

rotation $\theta(\lambda)$ oscillates in the vicinity of LSPR (**D**$_1$) and is same as that of Bi:YIG at LSPR (**D**$_2$) band. Moreover, for the SW range, a negligible change in **R** results in the drastic flip-over of the MO response at $\lambda = 630$ nm [see the left graph in Fig. 1(b), (c)]. Another interesting feature in the spectra is that the presence of the additional scatterers having the coordinates of (i**D**$_3x$, j**D**$_3y$) strongly influences the spectral position of LSPR (**D**$_1$). A small change in **R** results in the following spectral shifts: $\Delta\lambda$(LSPR(**D**$_1$)) ≈ 5 nm and $\Delta\lambda$ (LSPR (**D**$_2$))) ≈ 1 nm at Δ**R** = 1 nm. The observed peculiarities, result from the out-of-phase field oscillation on the neighbor nanoparticles in the case of LSPR (**D**$_1$). In contrast, there is an in-phase dipole-to-dipole interaction of the neighbor nanoparticles in the case of LSPR (**D**$_2$)–the likely reason why the MO response does not change sign. It is worth noting that in the vicinity of strong polarization rotation at 635 nm the light emerging from the Au–BiYIG layer is elliptically polarized, see Fig. 1 (b) and (c). As for polarization states of the waves at LSPR (**D**$_2$) and LSPR (**D**$_1$), they are linear ($\lambda = 733$ nm) and elliptical ($\lambda \approx 925$ nm) polarizations correspondingly, see Fig. 1 (c).

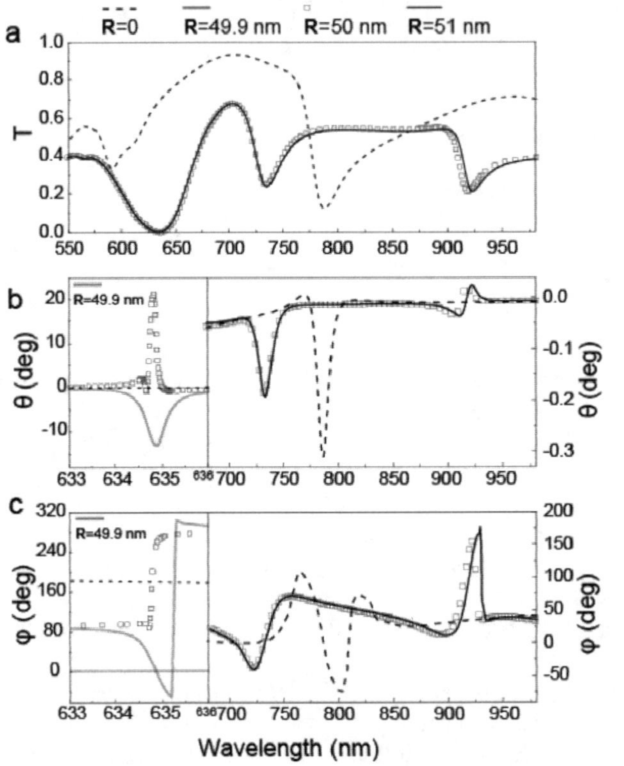

Fig. 1. a) Calculated transmittance (b) polarization rotation spectra and (b) phase shift between the **E**$_x$ and **E**$_y$ components for a nested 2D structure with **D**$_1$ = 350 nm. The pairs of spectra demonstrate their dependence on a negligible change of **R**. Spectra of the square 2D structure with **D**$_1$ = 350 nm are shown for comparison.

References

1. S. A Maier, *Plasmonics: Fundamentals and Applications* (Springer, USA, 2007)
2. A. V. Baryshev, H. Uchida, and M. Inoue, J. Opt. Soc. Am. B **30**, 2371 (2013)
3. A. V. Baryshev, A. M. Merzlikin, J. Opt. Soc. Am. B **33**, 1399 (2016)

Quantitative description of MLCT transitions in ruthenium polypyridyl complexes probed by two-photon absorption spectroscopy

Charles W. Stark[1], M. Rammo[1], K. Petritšenko[1], J. Pahapill[1], A. Mikhaylov[1], A. Rebane[1,2,*]

[1]Keemilise ja Bioloogilise Füüsika Instituut, Tallinn 12618, Estonia
[2]Department of Physics, Montana State University, Bozeman, USA

Abstract. Quantitative two-photon absorption (2PA) cross section and 2PA spectrum measurements were used to determine the molecular electric dipole change in the metal-to-ligand charge-transfer transition of ruthenium(II) tris-complexes of 2,2'-bipyridine ($[Ru(bpy)_3]^{2+}$) and 1,10-phenanthroline ($[Ru(phen)_3]^{2+}$) in several solvents.

1 Introduction

Ruthenium(II) polypyridyl complexes are well studied systems with the ability to form metal-to-ligand charge-transfer (MLCT) excited states with visible light absorption. This charge-transfer makes these complexes useful for numerous applications, but also makes them sensitive to local electric fields due to the formation of a dipole moment, $\Delta\mu$.
Quantitative two-photon absorption (2PA) spectroscopy is a versatile technique that can measure changes of permanent electric dipole moment.[1,2] We investigated the 2PA spectra of $[Ru(bpy)_3]^{2+}$ and $[Ru(phen)_3]^{2+}$ dissolved in water (H_2O), dimethylsulfoxide (DMSO), acetonitrile (ACN), and dichloromethane (DCM), and use a two-level model of 2PA applied to the lowest-energy transitions to estimate the value of $\Delta\mu$ in different solvents.

2 Results and Discussion

The 2PA cross-sections over the wavelength range of 680 – 1020 nm for $[Ru(bpy)_3]^{2+}$ and $[Ru(phen)_3]^{2+}$ are shown as solid lines in Figure 3. Applying the two-state model of 2PA[1] to the lowest-energy transitions, we can evaluate the difference in the ground- and excited state dipole moments according to the equation:

$$|\Delta\mu| = \frac{13650}{n^2+2}\sqrt{\frac{n}{\lambda}\frac{\sigma_{2PA}(2\lambda)}{\varepsilon_{1PA}(\lambda)}}, \quad (1)$$

where n is the index of refraction, ε_{1PA} is the molar absorption coefficient at wavelength λ, and σ_{2PA} is the 2PA cross section at twice that wavelength. The calculated $\Delta\mu$ values

* Corresponding author: arebane@montana.edu

(Table 1) are in the range 5 – 8 D and show a decrease with decreasing solvent polarity. Our all-optical measurements agree with previous electroabsorption measurements,[3,4] but are the first to show the dipole moment solvent dependence for both complexes, which is important in their function as a charge-transfer photosensitizers.

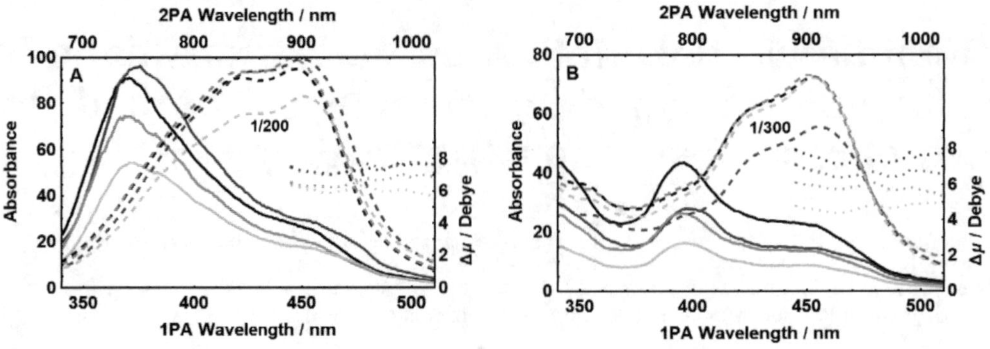

Fig 1. Absorption spectra collected for $[Ru(phen)_3]^{2+}$ (A) and $[Ru(bpy)_3]^{2+}$ (B) in H_2O (Black), DMSO (Dark-Gray), ACN (Mid-Gray), and DCM (Light-Gray). 2PA (solid lines) are in units of Goeppert-Mayer (GM), and 1PA molar absorption coefficients ($M^{-1}cm^{-1}$, dashed lines) are scaled for comparison. The calculated dipole moment ($\Delta\mu$) is displayed with units on the right-hand axis.

Table 1. Comparison of the 1PA molar absorption coefficient (ε) and 2PA cross section (σ) at λ_{max} for both complexes in the four solvents studied, as well as the relative permittivity (ε_r) for each solvent.

Solvent	ε_r	$[Ru(bpy)_3]^{2+}$			$[Ru(phen)_3]^{2+}$		
		ε (M^{-1} cm^{-1})	σ(GM)	$\|\Delta\mu\|$ (D)	ε (M^{-1} cm^{-1})	σ(GM)	$\|\Delta\mu\|$ (D)
H_2O	78	14450[a]	56	7.4	19000[a]	83	7.4
DMSO	47	11100[b]	32	6.6	19900[e]	93	7.4
CAN	36	14600[c]	27	6.0	19900[d]	73	6.6
DCM	9	14400[e]	26	4.5	16600[e]	54	6.0

[a]ref [5], [b]ref [6], [c]ref [7], [d]ref [8], [e]ref [9].

References

1. A. Rebane, G. Wicks, M. Drobizhev, T. Cooper, A. Trummal, and M. Uudsemaa, Angew. Chem. Int. Ed., **54**, 7582 (2015)
2. S. de Reguardati, J. Pahapill, A. Mikhailov, Y. Stepanenko, and A. Rebane, Opt. Express, **24**, 9053 (2016)
3. L. Karki, and J. T. Hupp, Inorg. Chem. **36**, 3318 (1997)
4. K. Kawamoto, Y. Tamiya, T. Storr, R.J. Cogdell, I. Kinoshita, and H. Hashimoto, J. Photochem. Photobiol. Chem. **353**, 618 (2018)
5. K. Kalyanasundaram, Coord. Chem. Rev. **46**, 159 (1982)
6. G. A. Heath, L. J. Yellowlees, and P. S. Braterman, J. Chem. Soc. - Chem. Commun., **0**, 287 (1981)
7. A. Yoshimura, M. Hoffman, and H. Sun, J. Photochem. Photobiol. Chem. **70**, 29 (1993)
8. R. J. Staniewicz, R. F. Sympson, and D. G. Hendricker, Inorg. Chem., **16**, 2166 (1977)
9. J. V. Caspar, and T. J. Meyer, Inorg. Chem., **22**, 2444 (1983)

Influence of solid matrices on spectral features of metalloporphyrins

Aleksandr Starukhin[1,*], *Aleksander* Gorski[2], and *Michal* Kijak[2]

[1]B.I. Stepanov Institute of Physics, National Academy of Sciences of Belarus, Minsk 220072, Belarus
[2]Institute of Physical Chemistry, Polish Academy of Sciences, Warsaw 01-224, Poland

Abstract. A way of simultaneously forming in the ground electronic state a planar, as well as two distorted forms was established for a set of metalloporphyrins in solid matrices at cryogenic temperatures. Experimental data together with results of theoretical DFT calculations are combined for analysis and assignment of possible types of distorted porphyrin macrocycles conformations.

Metallocompexes of porphyrins are biologically relevant compounds known for many years. On other side metalloporphyrins are suitable compounds for single molecule spectroscopy as molecules with wide variable different spectral features at cryogenics temperatures. However, despite intensive studies by many groups, their spectral features and photophysical properties are still the subject of long debates and new experiments. The spectroscopic studies at low temperatures belong to effective and proven methods of investigation of fine details such as a vibrational structure of molecules in both the ground and excited states. One of the most spectacular methods for recording fine-line luminescence spectra is the fluorescence line narrowing (FLN) [1].

In this report, we discuss the results, obtained for set of the simplest metalloporphyrins in different solid solutions, crystals and rare gas matrices at liquid helium temperature. The interest in such type of study was motivated by a unique possibility to obtain the spectra of the different spectral conformation of metalloporphyrins.

For example, fluorescence spectra of three different spectral forms Zn-porphine (ZnP), embedded in solid argon matrix are presented on Fig. 1. Fluorescence excitation spectra were recorded upon detection of fluorescence bands for corresponding spectral conformation also. These spectral features are caused by realization the structures with position of Zn- atom «in plane»» (Form F0 - Fig1a) and two types «out of plane» of porphyrin macrocycles in matrices (Form FI- Fig1b and FII- Fig1c) [2]. Above mentioned forms of ZnP in argon matrix have the different positions of the 0-0 transitions of the first electronic states at 552.3, 562.4 and 566.45 nm. The energy gap between positions of the 0-0 transitions more than 300 cm^{-1} of Form F0, for Form FI. Spectral gap beween Form F0 and Form FII have the value about 450 cm^{-1}. In a Soret band range the position of long-wave transitions for three forms are manifested at 385, 388 and 404 nm, correspondingly.

* Corresponding author: starukhin@mail.ru

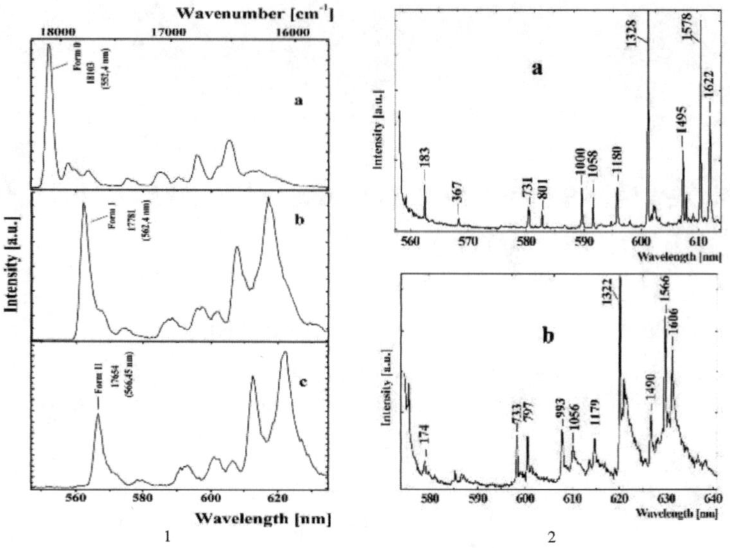

Fig. 1. Fluorescence spectra of ZnP (1) in solid argon matrix onto a sapphire substrate under nonselective excitation: $\lambda_{exc.}$ = 385(a), 388 (b), 404 nm (c) at T = 5 K and FLN spectra (2) of ZnP in *n*-octane, $\lambda_{exc.}$= 556,8 nm (a) and ZnP in *n*-octane with pyridine $\lambda_{exc.}$= 573,1 nm (b) at 4,2 K

Relative intensities of the 0-0 and vibronic transitions are very different for these forms (Fig. 1). Upon the fluorescence line narrowing conditions there are essential difference in intensities of vibronic transitions. It is connected with different character of vibronic borrowing intensities for vibrational transition upon formation of vibrational structure of fluorescence spectra for above-mentioned forms. FLN spectra of Form F0 (realized for ZnP in *n*-octane) and Form FII (ZnP in *n*-octane with pyridine) were recorded upon laser selective excitation. The modes with frequencies 1578 and 1622 cm^{-1} (Form F0) decrease to 1566 and 1606 cm^{-1}in spectrum of Form FII, correspondingly. An empirical interpretation of this effect can lead to wrong conclusions Density Functional Theory (DFT) and time dependent DFT methods were used for optimization of the ground and excited states together with calculation intensities and frequencies of normal modes of mettalocomplexes. Based on performed calculations, an assigmentof experimentally observed forms was proposed. The FI form was attributed to a complex of metalloporphyrin with one extra ligand placed above its plane. The FII form corresponds to a structure with two additional ligands located at opposite sides of the porphyrin macrocycle. Complexes of metalloporphyrins with rare gas may be realized as Van der Waals complexes between metalloporphyrin and rare gas. Upon condensation of the molecules into rare gas matrix, the distance between the matrix and the chromophore is less than one nanometer, which can lead to a significant increase of the probability of Van der Waals complexes formation [3].

This work was supported by the European Union's Horizon 2020 research and innovation program under grant agreement № 645628 and project № F18RA-012.

References

1. R. I. Personov, E. I. Al'Shits, L. A. Bykovskaya, Opt. Commun. **6**, 169 (1972)
2. A. S. Starukhin, A. M. Shul'ga, Opt. Spectrosc. **98**, 780 (2005)
3. U. Barash, V. L. Ginzburg, Physics-Uspekhi **143**, 345 (1984)

Optical detection of single transparent nanoparticles

Yuri Vainer[1,*], *Alexander* Zybin[2], *Nikolay* Gippius[3], and *Anastasia* Malek[4]

[1]Institute of Spectroscopy RAS, 142190 Moscow, Troitsk, Russia
[2]Leibniz-Institut für Analytische Wissenschaften, 44139 Dortmund, Germany
[3]Skolkovo Institute of Science and Technology, 121205 Moscow, Russia
[4]N.N. Petrov Institute of Oncology, 197758 Saint-Petersburg, Russia

Abstract. We will present the novel experimental far-field optical microscopy technique for detection of single nanoparticles, which weakly absorb a visible light and are characterized by refractive index close to its value in nearby environment (including particles of organic and biological origin: polymer nanoparticles, micro- and extracellular vesicles, liposomes, viruses etc).

Visualization of small nanoparticles is one of the basic challenges of the modern optical microscopy. Extremely low level of light signals makes this task very difficult. It is especially difficult to detect the so-called "invisible" nanoparticles, which weakly absorb a light of a visible band and are characterized by the index of refraction close to its value in nearby environment. Such objects are typical in biology media and are often used in modern technologies (for example, solution of polymer nanoparticles in water).

Despite the evident success of the fluorescence microscopy in high sensitive nanoparticle detection it has a set of fundamental shortcomings. First, a necessity of sample labelling that might alter properties of particles, and secondly, an undesirable photobleaching of fluorescence emitters and a fluorescence saturation, that limited the number of emitted photons.

We present the novel optical technique for high sensitive image of nearly invisible nanoparticles based an elastic scattering detection. Our approach uses the modified suffer plasmon assisted microscopy and waveguide technique that allows to increase significantly the efficiency of sample illumination. The original scheme of optical signal detection, sensitive to distortion of scattered light, allows to increase noticeably signal to noise ratio. The important advantage of our microscope is the possibility to detect the same particles simultaneously by two different methods. It can be realized by collecting light signals from two different sides of the sample. This way allows to raise the reliability of nanoparticle detection with the assumption that the origin of the experimental errors in two methods is different (see, e.g. Fig. 1 and Fig. 2).

* Corresponding author: vainer@isan.troitsk.ru

Fig. 1. Fragment of the detected image of polysterene nanospheres of 80 nm diameter.

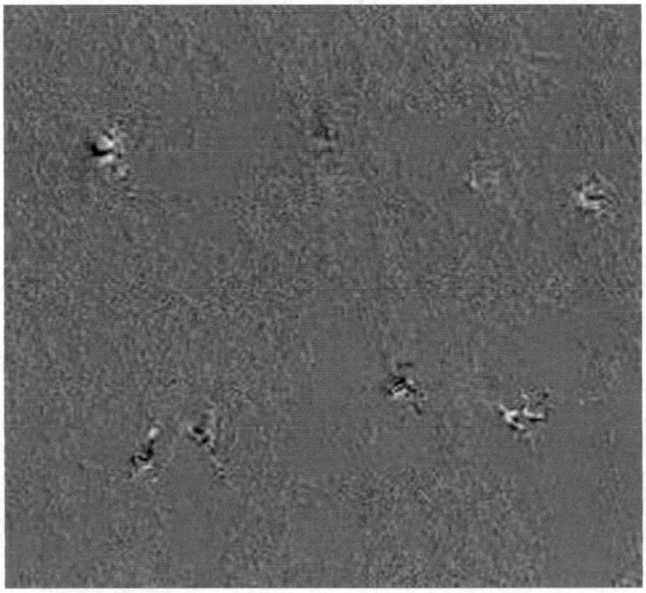

Fig. 2. Fragment of the detected image of liposomes of ~ 100 nm diameter.

Multifocus self-focusing of a femtosecond optical vortex

Evgeny Vasilyev*, *Svyatoslav* Shlenov, and *Valery* Kandidov

Lomonosov Moscow State University, Faculty of Physics and International Laser Center, 119991, Moscow, Russia

Abstract. We analyze formation of multifocus structure during the propagation of optical vortex with topological charge $m = 1$ in fused silica at wavelength 1800 nm, which corresponds to region of anomalous group velocity dispersion. Propagation of pulse is accompanied by sequence of nonlinear focuses, where each focus is a high intensive ring.

Self-action of femtosecond pulses may tend to formation of spatially long structures with high intensity – filaments [1]. Propagation of laser pulses in the regime of anomalous group velocity dispersion (GVD) may be accompanied by formation of localized in space and time high intensive structures – "light bullets" [2, 3]. Filamentation in beams with smooth phase has been widely studied for Gaussian beams, Airy and Bessel beams [4]. Annular beams with spiral phase dislocation have zero field on optical axis and are often called optical vortices. Self-focusing of optical vortex in a medium with normal GVD was studied in [5, 6]. Filamentation of vortices often tends to breaking up into several hot points due to modulational instability [7], but the distance of nonlinear focus can be less than the distance of modulational hot points formation. Thus, nonlinear focus preserves annular symmetry.

The formation of tubular filament with peak intensity 5×10^{13} W/cm^2 and plasma electrons concentration $10^{-3}N_0$ of neutrals, was demonstrated in [9] for femtosecond optical vortex in fused silica with normal GVD at central wavelength 800 nm. This work is devoted to numerical simulations of optical vortex filamentation in the presence of anomalous GVD. We analyze dynamics of light bullets and evolution of fluence distribution of the pulse.

Numerical simulation of the optical vortex self-action is based on the system of equations for slowly varying complex amplitude of laser field $A_V(r,\varphi,t,z) = A(r,t,z)e^{im\varphi}$ and free electron concentration $N_e(r,t)$:

$$\begin{cases} 2ik_0\dfrac{\partial A}{\partial z} = \hat{T}^{-1}\left(\dfrac{\partial^2}{\partial r^2} + \dfrac{1}{r}\dfrac{\partial}{\partial r} - \dfrac{m^2}{r^2}\right)A + \hat{T}^{-1}\hat{D}A + \dfrac{2k_0^2}{n_0}\hat{T}\Delta n_k A - \dfrac{2k_0^2}{n_0}\hat{T}^{-1}\Delta n_{pl}A + \\ \qquad + i\hat{T}^{-2}\sigma A - ik_0(\alpha + \delta)A, \\ \dfrac{\partial N_e}{\partial t} = W(N_0 - N_e) + N_e(\nu_i - \beta). \end{cases}$$

Operator of the wave nonstationarity $\hat{T}$ [9] is used to describe such effects as femtosecond pulse wavefront self-steepening.

* Corresponding author: vasilev.evgeniy@physics.msu.ru

Initial shape of optical vortex was considered as annular vortex beam with Gaussian envelope:

$$A_V(r,\varphi,t,z=0) = A_0\left(\frac{r}{r_0}\right)^m e^{-r^2/2r_0}e^{-t^2/2t_0}e^{im\varphi},$$

where $m = 1$ is topological charge, $\varphi = \arctan x/y$. Spatial parameter $r_0 = 100\ \mu$m, pulse duration $t_0 = 36$ fs, central wavelength $\lambda_0 = 1800$ nm. Peak power of initial beam was $P_0 = 5P_V$, where P_V is the critical power of self-focusing for optical vortex [5, 6] with topological charge $m = 1$.

In the initial stage of vortex beam propagation the self-action yields narrowing ring in the beam cross-section, its radius being practically the same $r_0 = 100\ \mu$m at $z = 0.9$ cm (Fig. 1). Peak intensity rises up to 1.5×10^{13} W/cm^2, ring width decreases in $\sim$10 times, peak fluence is 0.07 J/cm^2. This wave packet corresponds to the first nonlinear focus and can be considered as vortex bullet. Kerr's self-focusing is then stopped by linear diffraction, which is accompanied by flowing pulse energy towards the optical axis. Phase singularity prevents the appearance of the light field in the beam center and interference rings are formed at $z\sim1.2 - 2.5$ cm around the optical axis with fluence up to 0.04 J/cm^2. The next nonlinear focus takes place at distance $z = 4.1$ cm. The light bullet corresponding to the second focus has approximately the same peak intensity as in the first focus, ring radius being about 50 μm. Energy flowing towards beam axis remains up to the distance of $z\sim4.3$ cm, where the last focus is situated. Peak intensity reaches the global maximum, which is about 5×10^{13} W/cm^2, fluence is 0.24 J/cm^2, the ring radius decreases up to 10 μm. Plasma electrons concentration is reached for the first time noticeable value $0.5 \times 10^{-3}N_0$, where N_0 is neutrals concentration.

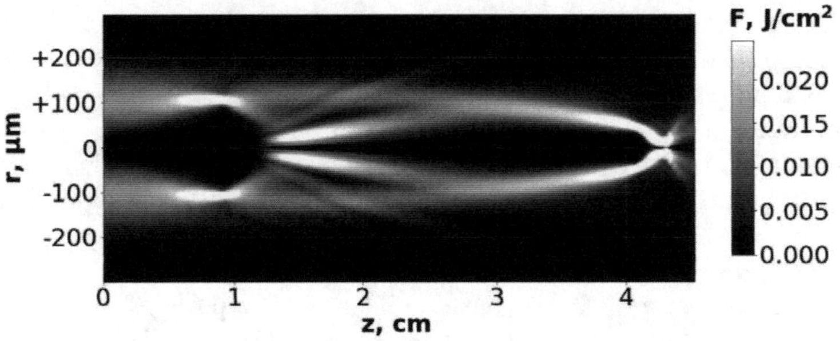

Fig. 1. Fluence evolution of optical vortex in fused silica at wavelength 1800 nm.

This research was supported by the RFBR grant 18-02-00624.

References

1. V. P. Kandidov, S. A. Shlenov et al. Quant. El., **39**, 205 (2009)
2. E. O. Smetanina, A.E. Dormidonov et al., Laser Phys., **22**, 1189 (2012)
3. S. V. Chekalin, A. E. Dokukina et al., J. Phys. B, **48**, 094008 (2015)
4. A. Couairon, A. Lotti et al., Proc. of SPIE, **8770**, 87701E (2013)
5. V. I. Kruglov, Yu. A. Logvin et al., J. Modern Optics, **39**, 2277 (1992)
6. L. T. Vuong, T. D. Grow et al., Phys. Rev. Let., **96**, 133901 (2006)
7. M. S. Bigelow, P. Zerom, R.W. Boyd, Phys. Rev. Lett., **92**, 083902 (2004)
8. R. A. Vlasov, V. M. Volkov, D. Yu. Dedkov, Quant. El., **43**, 157 (2013)
9. E. V. Vasil'ev, S. A. Shlenov, Quant. El., **46**, 1002 (2016)

EPJ Web of Conferences **190**, 03016 (2018) https://doi.org/10.1051/epjconf/201819003016
HBSM-2018

Multi-photon luminescence of single gold nanoparticles: Exploring the dynamics of plasmons and electron hole pairs

Frank Wackenhut and *Alfred* J. Meixner*

Institute of Physical and Theoretical Chemistry, Eberhard Karls University of Tübingen, Tübingen 72072, Germany

Abstract. We investigate the dynamics of the multi-photon excitation of single, isolated gold nanoparticles and show that the one/two/three emission can be actively switched by changing the excitation pulse duration.

Luminescence of bulk gold has first been reported in the pioneering experiments of Mooradian in 1969 [1] and was attributed to the radiative recombination of electron hole pairs. In the recent years it was pointed out that their optical properties, like elastic scattering [2] or one photon luminescence [3], are dictated by the particle plasmon.

Here we utilize multi-photon excitation to investigate the optical properties of single gold nanoparticles and report on efficient two and three photon luminescence emission based on the excitation of electron hole pairs. Furthermore, we show that the emission of the same gold nanoparticle strongly depends on the excitation conditions and can be actively switched by changing the excitation pulse duration from 500 fs (Fig.1(a)) to 100 fs (Fig.1(b)).

Fig. 1 (a) shows the emission spectrum of a single gold nanosphere exited with one (black line) and two (red line) photons exhibiting a single emission band at 550nm. This emission follows closely the plasmonic properties of the gold nanosphere. In Fig. 1(b) the gold nanosphere is excited with one (black line) and two (red line) photons and the green line displays a corresponding simulation. In this case a shorter excitation pulse duration was used, which drastically changes the emission properties. This emission can be described by the radiative recombination of electron hole pairs. Additionally, we are using single gold nanorods with different aspect ratios to investigate the interplay between the particle plasmon and electron hole pairs, which enables us to develop a quantitative model to fully describe the two and three photon luminescence emission of single gold nanoparticles.

These findings greatly increase the understanding of the physical processes underlying the luminescence of gold and hence help to increase the usefulness of gold nanoparticles in various application fields, e.g. in material science, bio imaging, microscopy and spectroscopy.

* Corresponding author: alfred.meixner@uni-tuebingen.de

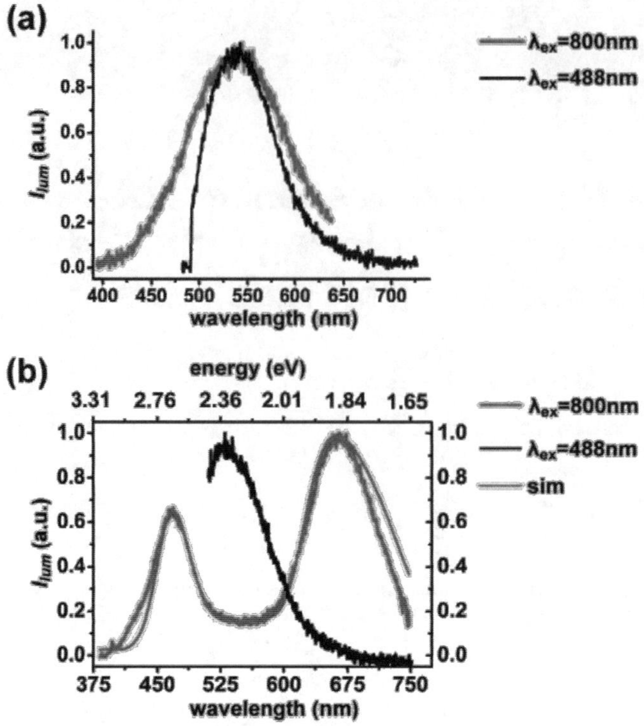

Fig. 1. (a) shows the typical plasmonic emission spectrum of a single gold nanosphere exited with one (black line) and two (red line) photons exhibiting a single emission band at 550nm. (b) displays the emission excited with one (black line) and two (red line) photons. In this case the pulse duration was reduced from 500 to 100fs causing a drastic change in the emission behaviour. The green line shows a corresponding simulated spectrum.

References

1. A. Mooradian, Phys. Rev. Lett., **22**, 185 (1969)
2. F. Wackenhut, A. Virgilio Failla, T. Züchner, M. Steiner, and A. J. Meixner, Appl. Phys. Lett. **100**, 263102 (2012)
3. F. Wackenhut, A. V. Failla, and A. J. Meixner, J. Phys. Chem. C, **117**, 17870 (2013)

EPJ Web of Conferences **190**, 04001 (2018) https://doi.org/10.1051/epjconf/201819004001
HBSM-2018

Spin colour centres in SiC as a material platform for sensing and information processing at ambient conditions

Andrey Anisimov[1,*], *Victor* Soltamov[1,2], *Pavel* Baranov[1], *Georgy* Astakhov[2,3] and *Vladimir* Dyakonov[2]

[1]A.F. Ioffe Institute RAS, Politekhnicheskaya 26, 194021, Saint Petersburg, Russia
[2]Experimental Physics VI, Julius-Maximilians University of Wuerzburg, 97074 Wuerzburg, Germany
[3]Helmholtz-Zentrum Dresden-Rossendorf, Institute of Ion Beam Physics and Materials Research, 01328 Dresden, Germany

Abstract. Atomic-scale colour centres in bulk and nanocrystalline SiC are promising systems for quantum photonics compatible with fiber optics, quantum information processing and sensing at ambient conditions. Colour centres which acts as stable single photon sources in SiC can be key elements for quantum photonics and communications. It has been shown that there are at least two families of colour centres in SiC with S = 1 and S = 3/2, which have the property of optical alignment of the spin levels even at room temperature and above. The spin state can be initialized, manipulated and readout by means of optically detected magnetic resonance (ODMR), level anticrossing and cross-relaxation. Recently, we observed the effects of “hole burning” in the ODMR spectra, which made it possible to narrow the ODMR line by approximately an order of magnitude, which substantially increases the possibilities of technological applications of spin centres.

The unique quantum properties of the NV centre in diamond [1, 2] have motivated efforts to find defects with similar properties in silicon carbide (SiC), which can extend the functionality of such systems. Atomic-scale colour centres in bulk and nanocrystalline SiC are promising for quantum information processing, photonics compatible with fiber optics and sensing at ambient conditions. Their spin state can be initialized, manipulated and readout by means of optically detected magnetic resonance (ODMR), level anticrossing (LAC), cross-relaxation (CR) [4] and hole-burning radio-frequency spectrum analysis [3]. It has been shown that there are at least two families of colour centres in SiC with S = 1 and S = 3/2, which have the property of optical alignment of the spin levels and allow a spin manipulation. For the S = 3/2 family, the ground state and the excited state were demonstrated to have spin S = 3/2 and a population inversion in the ground state can be generated using optical pumping, leading to stimulated microwave emission even at room temperature and above. By controlling the neutron irradiation fluence, the colour centres concentration can be varied over several orders of magnitude down to a single defect level.

* Corresponding author: aan0100@gmail.com

Furthermore, these atomic-scale spin colour centres can be also attractive for local or environment sensing. Several, separately addressable spin-3/2 centres (spin qudits) have been identified in the same crystal for each polytype (4H-, 6H- and 15R-SiC) which can be used either for magnetic field or temperature sensing. Some of these spin colour centres are characterized by nearly temperature independent zero-field splitting, making these centres very attractive for vector magnetometry. Contrarily, the zero-field splitting of the centres in the excited state exhibits a giant thermal shift, which can be used for thermometry applications. SiC is taking on a new role as a flexible and practical platform for harnessing the new quantum technologies, it is a technologically material, used in various devices (LED, MOSFETS, MEMS).

Here, we apply a two-frequency to excite and image multiple qudit modes in a SiC spin ensemble under ambient conditions. Strikingly, their spectral width is about one order of magnitude narrower than the inhomogeneous broadening of the corresponding spin resonance [3].

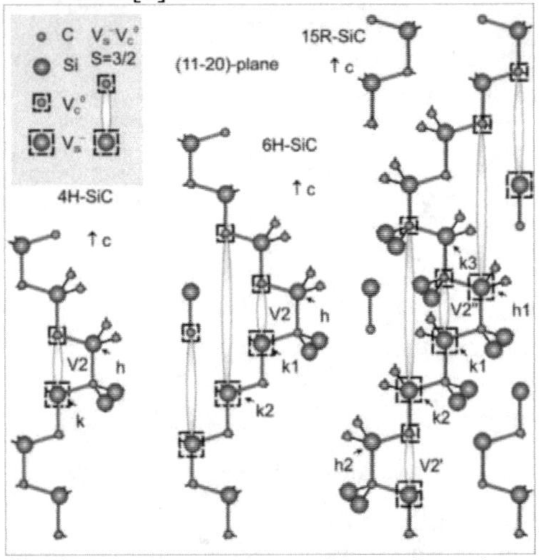

Fig. 1. Models showing possible configurations familie of spin colour centres in 4H-SiC, 6H-SiC and 15R-SiC lattice in (11-20) plane with S=3/2.

Figure 1 (a) show familie of spin colour centres in 4H-SiC, 6H-SiC and 15R-SiC with unique mechanism of an optical alignment of the spin sublevels. Each spin colour centres of is divided into several possible configurations, as there are different lattice positions in the crystal. Family of the centres, labeled as a V1, V2, V3 and V4 is a negatively charged silicon vacancy V_{Si}^{-} in the paramagnetic state that is noncovalently bonded to the neutral carbon vacancy V_C^0 in the nonparamagnetic state, located on the adjacent site along the SiC symmetry c axis having quadruplet ground and excited states (S=3/2). We apply a two-frequency to excite and image multiple qudit modes in a SiC spin ensemble under ambient conditions. Strikingly, their spectral width is about one order of magnitude narrower than the inhomogeneous broadening of the corresponding spin resonance.

We have presented a review of optical and spin properties of colour centres in different polytypes of SiC, which were studied two-radiofrequency hole-burning spectrum analysis. This study will allow SiC is taking on a new role as a flexible and practical platform for harnessing the new quantum technologies.

This work was supported by the by the RFBR under Grant № 16-02-00877-a; by the RSF № 16-42-01098 and V.A.S. acknowledges support through a sponsorship provided by the Alexander von Humboldt (AvH) foundation fellowship.

References

1. A. Gruber et al., Science **276**, 2012 (1997)
2. P. G. Baranov et. al., *Magnetic Resonance of Semiconductors and Their Nano-structures: Basic and Advanced Applications* (Springer, 2017)
3. V. A. Soltamov et al., arXiv:1807.10383
4. A. N. Anisimov et al., AIP Advances **8**, 085304 (2018)

Slow drift compensation in long-term measurements in fluorescence nanoscopy with blinking colloidal semiconductor quantum dots

Aleksei Baev[*] and *Aleksei* Rozhentsov

Volga State University of Technology, Yoshkar-Ola 420000, Russia

Abstract. In this work we present an approach for image processing accuracy increasing in fluorescence nanoscopy of single colloidal quantum dots with blinking fluorescent behavior. The approach is based on the estimation of parameters, describing the slow collective drift, according to the information about a long series of frames with fluorescent images of single blinking quantum dots. The results of collective drift compensation for single colloidal quantum dots CdSe / ZnS supper-resolution image are demonstrated.

The presence of a slow drift - spatial image shift in time - is one of the serious problems of various nanoscopy techniques. The characteristic values of the shifts can vary from tens to several hundred or more nanometers per hour. The potential reasons are instabilities in the sample positioning system, slow changes in the spatial position of the optical elements in the recording system associated with mechanical and temperature instabilities. As a result, the supper-resolution images of particles are "blurred", which leads to a decrease in the accuracy of the estimation of their parameters. In this paper, we propose an approach that reduces the effects of slow drift in long-term measurements in experiments on luminescent nanoscale with single flickering quantum dots. The approach is based on estimating the parameters of a function describing the collective drift and sub-pixel shift compensating in each CCD-camera frame in image post-processing.

In present time there exist a few approaches to compensate nanodrifts: by measuring the interframe shift based on calculating the cross-correlation function between frames [1], by using the special embedded markers [2], etc. In this paper, we propose an approach based on the use of information on the subdi-fractional coordinates of flickering quantum dots recorded in a sufficiently long series of frames (from several hundred to several thousand frames) that does not require special markers. This approach makes it possible to obtain drift-corrected superresolution images of single quantum dots without additional markers. To do this, information about the coordinates of all the QDs found in the series of frames under study is grouped into a common data array by subtracting the mathematical expectation of the coordinates of the QDs within the track from the coordinates of each QD. The set of points thus obtained is approximated by a polynomial of a given degree. The obtained functional dependence will characterize the law of global drift of all marks.

[*] Corresponding author: kruzenshteyn@yandex.ru

To show the possibilities of suggested approach we use the experimental data on fluorescence images of single colloidal QDs CdSe/ZnS (with diameter of emitting core ~4 nm) kindly provided by the Single-molecule research team from the Institute for Spectroscopy of the Russian Academy of Sciences [3-5]. The fluorescence of single QD were excited by CW laser (Coherent Verdi V6) on the wavelength 532 nm with excitation intensity ~100 W/cm^2. The fluorescence images were detected by EM CCD camera (Luca Andor) with exposure time 100 ms per frame.

Figure 1 demonstrates the dependence of the change in the subdiffractional coordinate Y for several single QDs and corresponding functions which describe observed global drifts for the coordinates X (red line) and Y (green line).

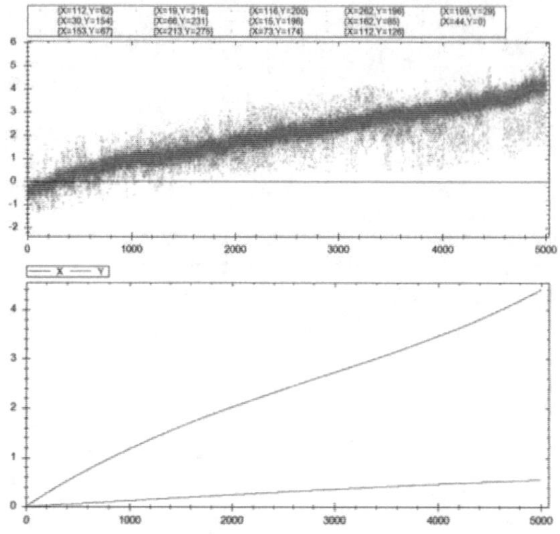

Fig. 1. Dependence of the change in the subdifraction coordinate Y for several quantum dots (a). The result of calculating the global drift for the coordinates X (red graph) and Y (blue graph) (b). The abscissa is the frame number.

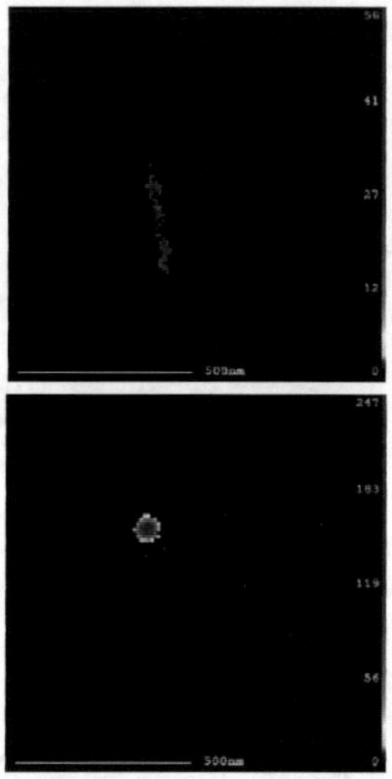

Fig. 2. Fragment of the histogram of the distribution of the single colloidal CdSe/ZnS QD coordinates, as restored from 5000 CCD-frames without (a) and with (b) compensation of the global drift.

The Figure 2 (a) shows an example of the “superresolution” image of single QD subjected to the slow drift. This superresolution image is constructed from restored (from each CCD-frame) QD coordinates as 2D-histogram of coordinates distribution. The significant drift along the Y coordinate is clearly observed. The “superresolution” image is elongated along the drift direction.

Figure 2b shows the image obtained for the same series of frames, after compensation for global drift. The density of the restored coordinates is a symmetrical figure, i.e., a circle of the radius equal to the accuracy of coordinates determination in a single measurement. Thus in this paper we show that slow drifts can be compensated by extended measurements and correct polynomial analysis.

The work was supported by the Russian Science Foundation (grant No. 17-72-20266).

References

1. M. J. Mlodzianoski et al., Opt. Express, **19**, 15009 (2011)
2. H. Ma et al, Biophysical Journal, **112**, 2196 (2017)
3. A. V. Naumov, I. Y. Eremchev, A. A. Gorshelev, EPJ D **68**, 348 (2014)
4. A. L. Shchukina, I. Y. Eremchev, A. V. Naumov, Phys. Rev. E, **92**, 032102 (2015)
5. I. Y. Eremchev, I. S. Osad'ko, A. V. Naumov, J. Phys. Chem. C, **120**, 22004 (2016)

Light shift of the Ramsey resonance of coherent population trapping in optically dense medium

Konstantin Barantsev*, *Gavriil* Voloshin, *Andrey* Litvinov, and *Evgeniy* Popov

Peter the Great St. Petersburg Polytechnic University, Department of theoretical physics, 195251 Saint-Petersburg, Polytechnicheskaia str., 29, Russia

Abstract. In this work we calculate the Ramsey resonances of coherent population trapping effect in cold atomic ensembles. The main goal is to investigate the light shift of the resonance on the different parameters in optically dense medium, because such medium can to distort and absorb the interrogating laser pulses.

Coherent population trapping (CPT) effect [1,2] allows us to detect the microwave transitions in atomic ensembles only by radiation of optical range. The two-frequency laser radiation interacts in simply case with the three-level Λ-system and induces the microwave atomic coherence. Atoms cease to scatter the light and the transparency window appears. Due to the fact that the microwave atomic coherence is long lived, the width of the transparency window is small compared with natural line width. It makes it possible to use this effect in wide range of practical applications such as quantum frequency standards [3], magnetometers [4], recording and processing of quantum information [5].

One of the ways to improve the accuracy of the metrological devices is to reduce the width of the reference resonance. Detection of the CPT resonance by the Ramsey method [6] allows us to further narrow the resonance line. The idea of this method consists of interrogation the atoms by the radiation not continuously, but in discrete periods. It can be done both by spatial separation of the interaction regions [7,8] and by using pulsed radiation [9-12]. The Ramsey method using pulsed radiation has been actively researched in the last years. One of the main problems affecting the accuracy of the atomic clock is the light shift of the reference resonance. At present, a new direction of research is the search for ways to suppress the light shifts of CPT-resonance when using the Ramsey method by varying of the shapes of the pulses, their phases and amplitudes – the so-called generalized auto-balanced Ramsey spectroscopy [13]. The work [14] uses a two-stage pulse shape to simultaneously suppress light shifts and improve the signal-to-noise ratio. In [15] the density matrix formalism is constructed for atoms in an arbitrary periodic field.

Despite the large number of modern works on the investigation of Ramsey detection method, there remain some open questions related to the influence of an optical density [16] of the medium at the Ramsey signal. The optically dense medium is the atomic medium where the mean free path of the quasiresonant photon is much smaller than the size of the medium. In such medium the laser radiation can be absorbed significantly, the form of the

* Corresponding author: kostmann@yandex.ru

interrogation laser pulses can be distorted and different modes are absorbed in different ways. This can drastically affect the total light shift of the CPT resonance, which is not taken into account in modern theoretical works in this direction. Moreover, the effects of dense medium can suggest new ways of suppressing the light shift of the Ramsey signal. The radiation modes can be equalized in intensity due to a different absorption coefficient, which will result in an auto-balancing effect. Such effect has already been observed by us in the excitation of the CPT resonance by continuous radiation [17].

In this work we investigate the influence of the optically dense medium on the light shift of the Ramsey resonance of coherent population trapping in cold atomic ensembles. The dependencies of the light shift on the intensity of radiation, durations of the laser pulses and dark pause, dipole moments of the atomic transitions, length of the medium and atomic concentration have been found.

This work is supported by Ministry of Education and Science of the Russian Federation in terms of FTP "Research and development on priority trends of Russian scientific-technological complex evolvement in 2014-2020 years (agreement 14.578.21.0211, agreement unique identifier RFMEFI57816X0211)".

References

1. G. Alzetta, A. Gozzini, L. Moi et al., Nuovo Cimento B, **36**, 5 (1976)
2. E. Arimondo, G. Orriols, Nuovo Cimento Lett., **17**, 333 (1976)
3. J. Vanier, Appl.Phys. B, 81, 421 (2005)
4. A. Akulshin, A. Celikov and V. Velichansky, Opt.Commun., **84**, 139 (1991)
5. M. Fleishhauer and M. D. Lukin, Phys. Rev. Lett., **84**, 5094 (2000)
6. N. F. Ramsey, Phys. Rev., **78**, 695 (1950)
7. J. E. Thomas, P. R. Hemmer, S. Ezekiel et al., Phys. Rev. Lett., **48**, 867 (1982)
8. I. M. Sokolov, Quantum Electronics, **45**, 947 (2015)
9. G. Pati, K. Salit, R. Tripathi et al., Opt. Commun., **281**, 4676 (2008)
10. Y. Yano, W. Gao, S. Goka, and M. Kajita, Phys. Rev. A, **90**, 013826 (2014)
11. G. S. Pati, Z. Warren, N. Yu, and M. S. Shahriar, JOSA B, **32**, 388 (2015)
12. P. Yun, F. Tricot, C. E. Calosso et al., Phys. Rev. Applied, **7**, 014018 (2017)
13. T. Zanon-Willette, V. I. Yudin and A. V. Taichenachev, Phys. Rev. A, **92**, 023416 (2015)
14. Y. Yano, S. Goka, M. Kajita, arXiv: 1701.05696v1 (2017)
15. V. I. Yudin, A. V. Taichenachev, M. Yu. Basalaev, Phys. Rev. A, **93**, 013820 (2016)
16. M. B. Gornyi, B. G. Matisov, Yu. V. Rozhdestvenskii, Zh. Eksp. Teor. Fiz., **95**, 1263 (1989)
17. K. A. Barantsev, A. N. Litvinov, E. N. Popov, I. M. Sokolov, IEEE Xplore, DOI: 10.1109/EFTF.2016.7477818 (2016)

Au/Ag-containing DNA-based nanowires

Dmitry Basmanov[1,*], *Kirill* Prusakov[1,2], *Dmitry* Klinov[1], *Gennady* Eidelstein[3], and *Aleksander* Kotlyar[3]

[1]Federal State Budgetary Institution Federal Research and Clinical Center Of Physical-Chemical Medicine Federal Medical Biological Agency, Laboratory of Medical Nanotechnologies, 119435 Malaya Pirogovskaya 1a, Moscow, Russia
[2]Moscow institute of physics and technology, Departure of Biological and Medical Physics, 141701 Institutskiy per. 9, Dolgoprudny, Moscow Region, Russia
[3]Department of Biochemistry and Molecular Biology, Faculty of Life Sciences and Center of Nanoscience and Nanotechnology, Tel Aviv University, Ramat Aviv, Tel Aviv 69978, Israel

Abstract. Direct conductivity measurements of thin uniform gold-coated DNA nanowires are presented. Gold-coated DNA conductive wires deposited on a mica surface are visualized by SEM and conductivity of the wires are measured using nanomanipulators built inside the microscope. The measurements show that the conductivity is limited by defects, and the thicker the coating the higher the conductivity of the wires. These gold-coated DNA nanowires are promising candidates for molecular electronics. DNA-base nanomaterial to be used in new DNA-based molecular electronics.

1 Introduction

DNA is considered to be one of the most promising candidates for future molecular electronics. Several types of DNA-based metalized nanowires have been recently reported. Among them are: new metalized form of DNA, E-DNA [1], and thin gold nanoparticles-coated DNA wires [2]. In this work we present direct conductivity measurements of such gold-coated DNA wires.

2 Materials and Methods

The suspension of gold seeds (diameter of about 1.5 nm) was poured on a mica substrate with deposited DNA and left on the surface for 4 h. The surface was then washed with ultrapure distilled water and dried by a nitrogen flow. The mica surface was subsequently flowed with a freshly premixed enhancement solutions of ascorbic acid. The ascorbic acid and the gold solutions were inserted into a mixing chamber by two peristaltic pumps, such that the mixture flushed the surface ≈1 s after the mixing. Treatment of the surface with the mixture was conducted for 10 s. The surface was then rinsed with water and dried with nitrogen gas [2].

* Corresponding author: dmitry.basmanov@niifhm.ru

This samples were visualized by Zeiss Merlin with GEMINI II Electron Optics SEM. The measurements of gold-coated DNA wires conductivity were performed with Kleindiek nanotechnik MM3A-EM micromanipulators mounted on an SEM adaptor plate inside the SEM vacuum chamber, and Keithley 2636A System SourceMeter (see Fig. 1).

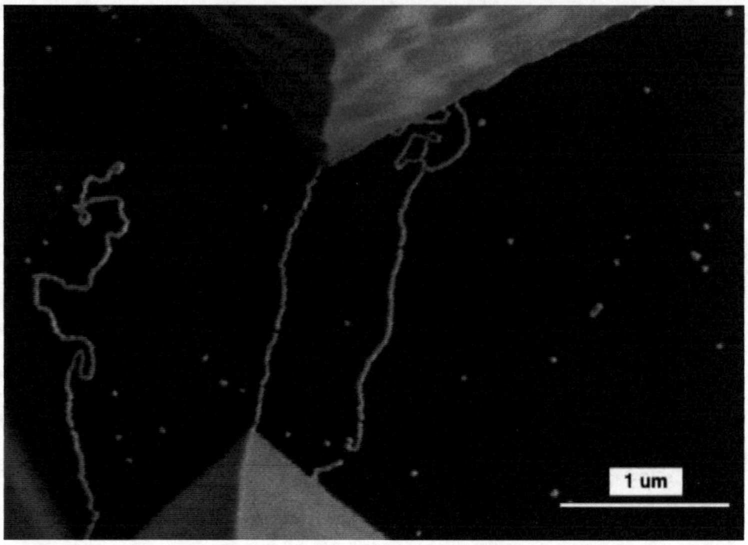

Fig. 1. Measurements of VAC of gold-coated DNA wires inside the Zeiss Merlin SEM.

3 Results and Conclusion

The thin gold-coated DNA wires display long-range conductivity. Our measurements show that the conductivity is limited by defects. Moreover, the thicker the gold coating the less the number of defects and the longer the conduction distance of the wire. The preparation method enables the formation of molecular wires with dimensions and uniformity that are suitable for molecular electronics.

This work was supported by the Russian Science Foundation (project № 17-75-30064).

References

1. G. Eidelshtein, N. Fardian-Melamed, V. Gutkin, D. Basmanov, D. Klinov, D.Rotem, Y. Levi-Kalisman, D. Porath, and A. Kotlyar, Adv. Mater, **28**, 4839 (2016)
2. A. Stern, G. Eidelshtein, R. Zhuravel, G. I. Livshits, D. Rotem, A. Kotlyar and D. Porath, Adv. Mater, **30**, 1800433 (2018)

Upconversion nanoparticles with anti-Stokes luminescence as bioimaging agents

Polina A. Demina[1,2,4,*], *Evgeniy* V. Khaydukov[2,3] , *Natalia* V. Sholina[3,5], *Vasilina* V. Rocheva[2], *Dmitry* A. Khochenkov[5], *Roman* A. Akasov[1,3] and *Alla* N. Generalova[1,2]

[1]Shemyakin-Ovchinnikov Institute of Bioorganic Chemistry RAS 117997, Moscow, Russia
[2]Scientific Research Centre "Crystallography and Photonics" RAS, 119333, Moscow, Russia
[3]I.M. Sechenov First Moscow State Medical University, 119991, Moscow, Russia
[4]Russian Technological University (MIREA), 119454, Moscow, Russia
[5]N. N. Blokhin National Medical Research Center of Oncology, 115478, Moscow, Russia

Abstract. Lanthanide-based upconversion nanoparticles attach great attention in theranostics due to their unique physicochemical and optical properties. It is innovative platform possessing peculiar properties for luminescent imaging, temperature mapping, sensing, and therapy. In present work we demonstrate advantages of new luminescent agents based on upconversion nanoparticles and hydrophylic biocompatible polymer.

1 Introduction

Theranoctics is a new interdisciplinary research area, which is an integration of therapy and diagnostics [1]. Upconversion nanoparticles (UCNPs) are promising nanoplatform for theranostics.[2] Lanthanide-based UCNPs are composed of inorganic crystalline ceramic host matrix, co-doped with pairs of different trivalent lanthanide ions (Ln^{3+}), usually the Yb^{3+}, Er^{3+} or Tm^{3+}. Its distinctive feature is the ability to convert near-infrared (NIR) light into higher energy visible or ultraviolet (UV) light [3]. UCNPs have been successfully used for in vivo bioimaging due to the low biotissue autofluorescence. The most common mechanism for passive nanoparticle delivery into solid tumors is the enhanced permeability and retention effect (EPR) allowing UCNPs to preferentially diffuse and accumulate in tumor tissues [4]. The efficiency of the EPR-effect in UCNP delivery into the tumor is associated with the circulation half-lifetime in blood.

In present work we developed nanocomplexes with prolonged circulation time in the blood at intravenous injection by UCNP surface modification with colominic acid (CA). CA is endogenous non-immunogenic, biodegradable, non-toxic compound with low serum protein adsorption [5]. These properties determine the prospects of using for solid tumour visualization in vivo. We studied chemical and photophysical properties of CA-modified UCNP agents and demonstrated its effective accumulation in tumour due to prolonged blood circulation time for 3 hours in comparison with UCNPs-polyethylene glycol complexes.

* Corresponding author: polidemina1207@yandex.ru

2 Results

Nanoparticles are synthesized in organic solvent that determines their hydrophobic surface properties. The development of hydrophilization methods for obtaining hydrophilic UCNP probes, which can be homogeneously dispersed in an aqueous medium, biocompatible and colloidal stable in a biological environment is crucial. Firstly, we hydrophilized UCNPs through coating nanoparticles with polyethyleneimine (PEI) having NH2 groups, without removing the oleic acid from the UCNP surface. Then, we modified the surface in two different ways: 1) by forming a covalent bond between the PEI and the CA using the carbodiimide activation method (CA-CDI); 2) by adsorption of the CA on the UCNP surface due to electrostatic interactions (CA(+/-)). The obtained samples are hydrophilic retaining their colloidal stability in PBS, pH=7.2 for 1 month. IR spectroscopy successfully confirmed the modification.

The minimal adsorption of blood proteins on the particles provided the prolonged particle circulation time in blood and effective accumulation in the tumor. Our study showed that after incubation CA-modified UCNPs with mouse serum at T=37°C for 1 hour the probes practically did not adsorb the serum proteins as compared to UCNPs/PEI. These data were obtained by detection the protein concentration in the supernatants using the Bradford method. Thus, CA-modified UCNPs have potential to prolonged circulation in the body.

The UCNPs accumulation in cells was studied by flow cytometry on the cell lines of murine macrophages RAW 264.7 and human monocytes THP-1. The cell suspension (106 cells per ml) was incubated with 0.1 mg/ml modified UCNPs for 30 min in a CO_2 incubator (37 C, 5% CO_2). UCNPs covered with CA-CDI showed a higher level of accumulation in the cells than CA(+/-).It is worth noting that the level of CA(+/-)-UCNP accumulation was comparable to the accumulation of PEG-modified UCNPs.

We studied in vivo circulation time in blood after intravenously injection in mice Balb/c and collected blood samples from the tail vein at several time intervals. The samples were analyzed using home built anti-Stokes fluorescent microscope [6]. Images from four random areas were taken and calculation of UCNP number in all areas was made.

CA coating of UCNPs made it possible to reduce the UCNP interaction with serum proteins, leading to longer circulation in the bloodstream of small animals (up to 3 hours) as compared to UCNPs, modified with PEG, which are in the bloodstream no longer than 1 hour.

The reported study was partly funded by RFBR according to the research project No. 17-03-01033.

References

1. X. Chen, S. S. Gambhir, J. Cheon Acc. Chem. Res. **44**, 841 (2011)
2. G. Chen, H. Qiu, P. N. Prasad, X. Chen, Chem. Rev. **114**, 5161 (2014)
3. A. N. Generalova, B. N. Chichkov, E. V. Khaydukov, Adv. Colloid Interface Sci. **245**, 1 (2017)
4. J. Fang, H. Nakamura, H. Maeda, Adv. Drug Deliv. Rev. **63**, 136 (2011)
5. A. I. Fernandes, G. Gregoriadis, Biochim. Biophys. Acta **1341**, 26 (1997)
6. A. N. Generalova, V. V. Rocheva, A. V. Nechaev, D. A. Khochenkov, N. V. Sholina, V. A. Semchishen et al., RSC Adv. **6**, 30089 (2016)

Models of semiconductor quantum dots blinking based on the spectral diffusion

Vladislav Busov and *Pavel* Frantsuzov*

Lomonosov Moscow State University, 119991 Moscow, Russia

Two decades have passed since the first observation of long-term fluorescence intensity fluctuations (blinking) of single colloidal CdSe quantum dots (QDs) with a ZnS shell [1]. In further experimental studies it was found (see [2-4]) that these fluctuations have a wide spectrum of characteristic timescales, from hundreds of microseconds to hours.

Another interesting phenomenon manifested in the emission of single quantum dots is spectral diffusion showing the characteristic time scales in order of hundreds of seconds [5-6]. It is not surprising that there are a number of models connecting the fluctuations of the emission intensity with the slow motion of the exciton energy. The first model of that kind suggested by Shimizu et al. [7] is based on the Efros/Rosen *charging mechanism* [8]. More advanced version of this idea was suggested by Tang and Marcus [9-10] as the diffusion-controlled electron transfer model (DCET). In 2014 Zhu and Marcus [11] presented an extension of the DCET model by introduction of an additional biexciton charging channel. Simultaneously with the Tang and Marcus, the model based on the alternative *fluctuating rate mechanism* of blinking was suggested by Frantsuzov and Marcus [12]. The mechanism assumes that the variations of the non-radiative exciton recombination rate k_n generate fluctuations of the emission intensity on a long time scale. The Frantsuzov and Marcus model connects the recombination rate with the fluctuating energy difference between 1S and 1P states.

The predictions of the DCET model [9-10], the Expend DCET model [11], and the Frantsuzov and Marcus model [12] were compared by the analytical analysis and numerical simulations. It was found that the DCET model [9-10] predicts the appearance of extremely long ON and OFF time periods in a single QD emission trace. As seen on Fig.1 such a period could last years, which is much longer than the duration of a typical experiment. Thus the QD can become permanently bright or permanently dark after about a hundred blinking cycles with a high probability. The presence of a second ionization channel in the Extended DCET [11] model resolves the problem with very long ON times, but not with the very long OFF times. As a result, the most of the QD has to become permanently dark in this model as confirmed by numerical simulations. These predictions significantly differ from the experimentally observed behaviour of single quantum dots. On the contrary, the predictions of the model of Frantsuzov and Marcus [12] correspond to the basic properties of the quantum dots blinking obtained in experiment, such as a nearly continuous set of emission intensities,

* Corresponding author: pavel.frantsuzov@gmail.com

the power-law form of the ON-time and OFF-time distributions obtained by the threshold procedure, and a change in behavior the of the emission intensity power spectral density on millisecond time scales.

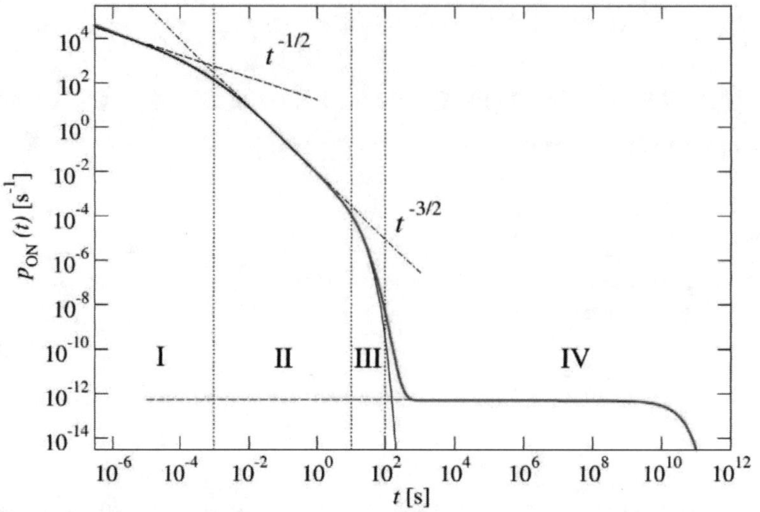

Fig. 1. The ON time distribution function within the DCET model (thick red line), the first interval power law (black dashed line), the second interval power law (black dashed-dotted line), the Tang-Marcus approximation (thin black line) and the long-time exponential asymptotic (dashed red line).

The study was supported by the Russian Foundation for Basic Research, project 16-02-00713.

References

1. M. Nirmal, B. O. Dabousi, M. G. Bawendi, J. J. Maklin, J. K. Trautman, T. D. Harris, and L. E. Brus, Nature, **383**, 802 (1996)
2. P. A. Frantsuzov, M. Kuno, B. Janko, and R. A. Marcus, Nature Phys., **4**, 519 (2008)
3. F. Cichos, C. von Borczyskowski, and M. Orrit, Curr. Opin. Colloid. Interface Sci., **12**, 272 (2007)
4. J. A. Cordones and S. R. Leone, Chem. Soc. Rev., **42**, 3209 (2013)
5. S. A. Empedocles, D. J. Norris, and M. G. Bawendi, Phys. Rev. Lett., **77**, 3873 (1996)
6. S. A. Empedocles and M. G. Bawendi, J. Phys. Chem. B, **103**, 1826 (1999)
7. K. T. Shimizu, R. G. Neuhauser, C. A. Leatherdale, S. A. Empedocles, W. K.Woo, and M. G. Bawendi, Phys. Rev. B, **63**, 205316 (2001)
8. A. L. Efros and M. Rosen, Phys. Rev. Lett., **78**, 1110 (1997)
9. J. Tang and R. A. Marcus, J. Chem. Phys., **123**, 054704 (2005)
10. J. Tang and R. A. Marcus, Phys. Rev. Lett., **95**, 107401 (2005)
11. Z. Zhu and R. A. Marcus, Phys. Chem. Chem. Phys., **16**, 25694 (2014)
12. P. A. Frantsuzov and R. A. Marcus, Phys. Rev. B, **72**, 155321 (2005)

EPJ Web of Conferences **190**, 04007 (2018) https://doi.org/10.1051/epjconf/201819004007
HBSM-2018

The study of a new family of phase masks for three-dimensional fluorescence nanoscopy

Aleksei Gorshelev[1,*], *Ivan* Eremchev[1,2], *Sergei* Kulik[1], *Andrei* Naumov[1,2,3], *Evgeniy* Vorontsov[4], *Vladimir* Volostnikov[4], and *Svetlana* Kotova[4]

[1]Institute of Spectroscopy RAS, 108840 Troitsk, Moscow, Russia
[2]Moscow Institute of Physics and Technology, 141700 Dolgoprudny, Moscow region, Russia
[3]Moscow State Pedagogical University, 119991 Moscow, Russia
[4]Lebedev Physical Institute, 443011, Samara, Russia

Abstract. The question of the axial coordinate determination accuracy is quite sophisticated in far field 3D fluorescence nanoscopy. The accuracy of a point emitter axial coordinate reconstruction depends among other things on the conversion efficiency of the phase mask and the microscope objective instrumental function. It was found that the instrumental functions of different microscope objectives differ significantly from each other, most have a strongly non-uniform spatial distribution of the radiation intensity in a parallel beam created by an objective focused on a point emitter. It was shown that taking into account the actual microscope objective instrumental function when calculating the phase masks allows to increase significantly the axial coordinate reconstruction accuracy.

Fluorescence microscopy with ultra-high spatial resolution (far field fluorescence nanoscopy, FFN) is currently one of the most effective and in-demand tools for research in the fields of biological and medical physics, biology, and chemical physics. It is also of interest for applications in solid state physics. Many of these far field fluorescence nanoscopy techniques (STORM, PALM, etc.) are based on the possibility of fluorescence imaging of single point-like emitting labels with nanoscale reconstruction of their lateral spatial coordinates, and subsequent computer rendering of object nanostructure by mapping plenty single labels positions (see [1,2] and references therein). Application of the technique based on the theory of spiral beam optics allows to reconstruct all three spatial coordinates of a single point emitter with nanometer accuracy. In this technique the spatial phase modulation of the light beam is used for the transformation of the single emitter point spread function into double helix point spread function [3].

The accuracy of a point emitter axial coordinate reconstruction depends on a variety of experiment parameters: the number of recorded photons, the conversion efficiency of a phase mask, the microscope objective instrumental function, and the characteristics of the optical scheme used in the registration channel.

FFN methods can be especially informative if fluorescent single molecules (SM) of organic dyes are used as emitting point light sources at cryogenic temperatures, when zero-

* Corresponding author: gorshelev@isan.troitsk.ru

phonon lines (ZPL) which correspond to purely electronic transitions in SMs are reachable for the observation. First, the analysis of ZPL of single molecules makes it possible to study the dynamic processes in impurity solid media, since the ZPL parameters are sensitive to the SM local environment characteristics. Second, the very large ratio of inhomogeneous absorption bandwidth to the very narrow ZPL homogeneous spectral width gives the opportunity to measure the fluorescence excitation spectra and coordinates for a huge number (up to all) efficiently fluorescing SMs in doped bulk solid samples - crystals, glasses, polymers (see [1,2] and references therein).

To perform measurements at cryogenic temperatures, it is necessary to use special micro-objectives designed for the operation at liquid-helium temperature, including in superfluid helium. In the case of an ideal microscope objective the propagation of light from a single point emitter located in a focal plane will be described after the objective by a plane wave with a Gaussian spatial intensity profile. However, in reality there are a variety of aberrations that degrade the picture. Manufacturers of the microscope objectives try to compensate for certain aberrations, depending on the tasks for which an objective is calculated. As a result, in most cases the spatial distribution of the radiation intensity in a parallel beam created by an objective focused on a point emitter, - the microscope objective instrumental function, - is significantly different from the ideal one. In particular, it is peculiar to micro-objectives designed to operate at cryogenic temperatures.

In the case of such micro-objectives the use of phase masks calculated for the spatial distribution of the radiation intensity with a flat (or Gaussian) profile gives a low efficiency of conversion of the microscope objective instrumental function to the two-lobe function. In this connection the actual task is the calculation of new effective phase masks taking into account the shape of the microscope objective instrumental function.

In the present work with the use of single-crystal semiconductor colloidal quantum dots (QDs) as point emitters of radiation the instrumental functions of a number of micro-objectives were measured and analysed. It was found that the instrumental functions of different micro-objectives differ significantly from each other, most have a strongly non-uniform spatial distribution of the radiation intensity in a parallel beam. With the use of an objective with the radiation intensity spatial distribution profile closest to a flat one, for a number of original calculated phase masks [4] the efficiency of the microscope objective instrumental function to the two-lobe function conversion was estimated. It was found that taking into account the actual microscope objective instrumental function when calculating the phase masks allows to increase significantly the conversion efficiency.

It was also shown that during the calculation of the optimal phase mask, it is necessary to take into account the actual diameter of the light beam and the pixel size of the phase modulator. Calibration curves – the correspondence between the two-lobe function rotation angle and the position of the radiating QD along the optical axis (axial coordinate) - were measured for a number of phase masks using specially developed technique. In addition, the question of the accuracy of the QD axial coordinate reconstruction was experimentally investigated.

This work was financially supported by RFBR (project 16-29-11809).

References

1. A. V. Naumov, Physics Uspekhi, **56**, 605 (2013)
2. A. V. Naumov, I. Yu. Eremchev, A. A. Gorshelev, Europ. Phys. J. D, **68**, 348 (2014)
3. S. R. P. Pavani, R. Piestun, Opt. Express, **16**, 3484 (2008)
4. E. N. Vorontsov, S. P. Kotova, N. N. Losevsky, D. V. Prokopova, S. A. Samagin, Bull. Lebedev Phys. Inst., **45**, 71 (2018)

Low-temperature dynamics in a dye-doped polymer: correspondence between the data obtained by photon echo and single molecule spectroscopy

Kamil Karimullin[1,2,*] and *Andrei* Naumov[1,2]

[1]Institute for Spectroscopy, 108840 Troitsk, Moscow, Russia.
[2]Moscow State Pedagogical University, 119435 Moscow, Russia

Abstract. Low temperature dynamics (tunneling and vibrational relaxation) in doped polyisobutylene film has been reinvestigated using 2-pulse incoherent photon echo (2IPE) and compared with single-molecule spectroscopy (SMS) data. It has been shown that in a very wide range of low temperatures the 2IPE gives optical dephasing times which correspond to the narrowest zero-phonon lines of single dye molecules.

Combined photon echo (PE) and single-molecule spectroscopy (SMS) technique is a very powerful tool for detailed study of low-temperature dynamics in dye-doped disordered polymers and glasses [1]. However this type of studies reveal discrepancies when comparing dephasing times T_2 and single-molecule (SM) zero-phonon spectral lines (ZPL) width obtained from PE and SMS, correspondingly. For tetra-tert-butylterrylene (TBT) in polyisobutylene (PIB) PE-experiments obtained T_2-times much shorter than it is sufficient to cause the corresponding ZPL broadening in the temperature range of a few – dozen of Kelvins [2, 3]. We experimentally solved this contradiction in Ref. [4]. In this paper we show that in temperature range from 4.5 up to 20 K the 2-pulse incoherent PE (2IPE) technique gives T_2-times which correspond the narrowest SM ZPL.

The new series of PE-measurements were performed using the home-build 2IPE spectrometer that was constructed originally in the mid of 1990ies [1, 5], and recently reconstructed to achieve modern experimental possibilities [4, 6]. We used a thin (0.5 mm) film of PIB with TBT molecules. PIB ($M_w = 4.2\times10^6$ g·mol^{-1}, Sigma Aldrich) was dissolved in spectrally pure toluene (Sigma Aldrich) together with TBT. The sample was prepared by disposing of the solution between two glass slides and further pumping for toluene removal.

We measured 2-pulse IPE (2IPE) decay curves at different cryogenic temperature, and obtained the temperature dependence of the inverse optical dephasing time $1/\pi T_2$. This dependence is shown in Fig. 1 (large solid circles) together with previous 2IPE-data as measured originally in [7] (solid squares); T-dependences of ZPL for several single TBT molecules in PIB (small colour dots), as measured in [8]; and T-dependence of average spectral ZPL widths for a number of TBT SM in PIB (blue stars) as measured in [9].

In both experiments [8, 9] the fluorescence excitation spectra of single TBT molecules in thin (0.5 – 1 μm) PIB film were measured by tuning of frequency of narrowband (1 –

* Corresponding author: kamil@isan.troitsk.ru

2 MHz) dye CW-laser and detection of Stokes-shifted luminescence. In Ref. [8] the confocal scheme together with avalanche photodiode (time of measurement for one spectrum was t_m=250 s), whereas in Ref. [9] the wide-field luminescence microscope was used with detection by EM CCD camera (t_m ~ 10 s).

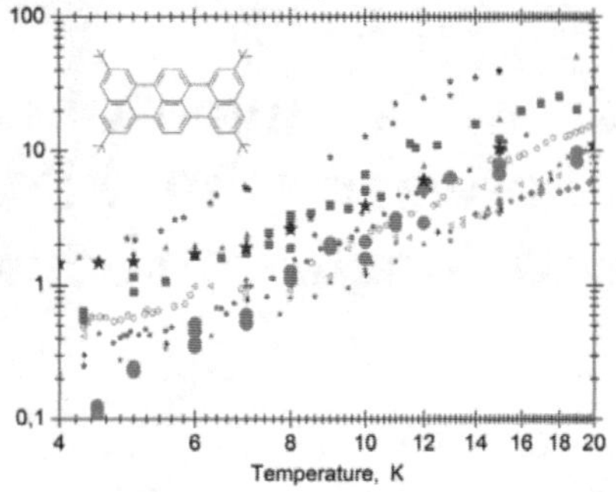

Fig. 1. Temperature dependences of ZPL homogeneous width measured by means of 2IPE (solid circles – new data, solid squares – previous data) and SMS (small dots and stars).

While the effective t_m for 2IPE (~ 10 ns) is much less than the effective t_m in SMS-measurements (~ 250 s), the average value of $1/\pi T_2$ should be in the region of narrowest SM ZPL widths. It is exactly what was observed in the comparative SMS/PE-studies at T = 2 K [8]. At higher temperatures (4–20 K) the ZPL broadening (as well as optical dephasing) is mainly caused by electron-phonon coupling, although the spectral diffusion processes are also intensified with *T*. Again, the value of $1/\pi T_2$ should be in the region of the narrowest ZPL widths. However, this was not observed in previous studies [1, 7, 8] for a broad range of temperatures. If we compare previous PE data with SM and average SM ZPL widths, we see that $1/\pi T_2$ comparable or even larger than average SM ZPL widths, that has not reasonable explanation.

We have carefully re-measured dephasing times for TBT in PIB in a very wide range of cryogenic temperatures from 4.5 K up to 20 K. The high sensitivity of the updated 2IPE-setup allows one to eliminate any saturation and light induced effects. As one can see from the Fig. 1 the new 2IPE data are in reasonable agreement with theoretical considerations. The $1/\pi T_2$ values correspond to the narrow ZPLs.

As possible explanations for the question why the previous PE-data did not agree with SMS, we can suggest the following two: (1) The additional contributions into the effective dephasing processes were observed to be due to high laser excitation power. Now the modern detectors and adjustment optical scheme [6] gives opportunity to diminish the laser pulse energy enough to obtain unsaturated values of $1/\pi T_2$. (2) Second possible reason is difference of molecular weight of PIB used in PE- and SMS-data. As it is known now [10] when diminish the M_w, additional spectral diffusion processes are strongly activated. As information about molecular weight of PIB is absent in first original PE researches, one can assume, that it was PIB with lower M_w in comparison with SMS-experiments, that leads to larger values of $1/\pi T_2$ measured by IPE. In this case the main role plays the spectral diffusion processes in nanosecond time scale.

The work was supported by the RFBR (project # 18-02-01121).

References

1. K.R. Karimullin, A.V. Naumov, J. Luminescence, **152**, 15 (2014)
2. A.V. Naumov et al., Phys. Stat. Sol. B, **241**, 3480 (2004)
3. A.V. Naumov et al., Phys. Stat. Sol. B, **241**, 3487 (2004)
4. M.V. Knyazev, K. Karimullin, A. Naumov, Phys. Stat. Sol. RRL, **11**, 1600414 (2017)
5. Y.G. Vainer et al., J. Luminescence, **86**, 265 (2000)
6. K.R. Karimullin et al., Meas. Science and Technology, **24**, 027002 (2013)
7. S.J. Zilker et al., J. Luminescence, **76-77**, 157 (1998)
8. Y.G. Vainer, A.V. Naumov, M. Bauer, L. Kador, J. Chem. Phys., **122**, 244705 (2005)
9. A.A. Gorshelev et al., Chem. Phys. Chem., **11**, 182 (2010)
10. I.Y. Eremchev et al., Phys. Chem. Chem. Phys., **13**, 1843 (2011)

Superconducting single-photon detector for integrated waveguide spectrometer

Alexander Korneev[1,2,*], *Vadim* Kovalyuk[1], *Pavel* An[1], *Alexander* Golikov[1], *Evgeniya* Zubkova[2,1], *Simone* Ferrari[3], *Oliver* Kahl[3], *Wolfram* Pernice[3], and *Gregory* Goltsman[1,2]

[1]Moscow State Pedagogical University, 119991 Moscow, Russia
[2]National Research University Higher School of Economics, 101000 Moscow, Russia
[3]University of Münster, Institute of Physics, 48149 Münster, Germany

Abstract. We present our recent achievements in the development of an on-chip spectrometer consisting of arrayed waveguide grating made of Si_3N_4 waveguides and NbN superconducting single-photon detector.

Superconducting single-photon detectors (SSPD) [1,2] became recently a device of choice in many applications ranging from quantum dots characterization to quantum cryptography and long distance space communication. The key factor of their success is superior performance: above 90% detection efficiency, very low dark count rate, picosecond timing jitter and high photon count rate. The superb performance makes SSPD especially attractive for advanced applications in quantum optics.

At the same time, there is a long-term trend in practical quantum optics to switch to optical integrated circuits on a single chip in which the light is delivered from the source to the detector by the dielectric waveguides. The waveguides were realized by means of several materials platforms, such as silicon on insulator (SOI), gallium arsenide (GaAs), silicon nitride (Si_3N_4) and polycrystalline diamond.

In [3] integration of SSPD with dielectric optical waveguides was demonstrated featuring near unity absorption for the wave travelling in the waveguide. At the same time, due to shorter nanostrip length compared to the traditional meander-shaped SSPD it features also shorter dead time and thus is capable of higher photon counting rates. Advanced photon detection capabilities offered by SSPD make such integrated circuits feasible for applications in optical signal processing, optical sensing and linear optical quantum computing.

In this paper, we present an approach to the realization of an on-chip spectrometer based arrayed waveguide grating (AWG) demultiplexer integrated with SSPD as shown in Fig. 1(a). The AWG is patterned from silicon nitride Si_3N_4 on silicon substrate with buried oxide by e-beam lithography. The input light is split by the AWG in several output channels, each with its own wavelength. We fabricated two types of eight-channel devices operating near the wavelength of 1550 nm and 740 nm. The main characteristics of the devices, including the free spectral range and channel width are presented in Table 1.

Each output waveguide is integrated with the U-shaped SSPD. In order to realize

* Corresponding author: akorneev@rplab.ru

optimal detector performance, the detector layout which produced the best results at 1550 nm during the previous characterization was chosen. All integrated SNSPDs on the AWGs were therefore 80 nm wide, 80 μm long with a gap of 100 nm between the stripes. The whole device is operated at 1.7 K temperature providing best performance of the SSPD. Fig. 1(b) shows on-chip detection efficiency (probability to detect photons of a wave in the waveguide, the losses at the input couplers are not taken into account) and dark count rate of the SSPD as a function of its bias current normalized to the critical current at which superconducting state is destroyed.

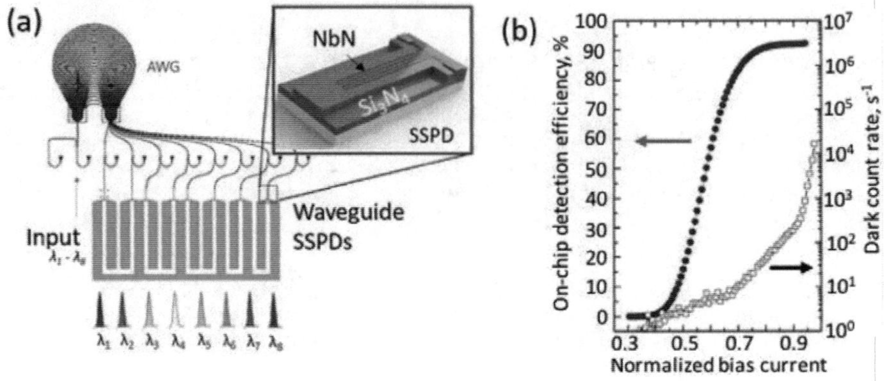

Fig. 1. (a) Schematics of AWG demultiplexer integrated with SSPD; (b) on-chip detection efficiency and dark counts rate of SSPD as functions of SSPD normalized bias current, a ratio of bias current I_b to the critical current I_c. Maximum on-chip detection efficiency can be as high as 92% at below 100 dark counts per second.

Table 1. AWG Parameters for the spectrometer.

Central wavelength, nm	Diffraction order	Free spectral range (FSR), nm	Transmission window, nm	Channel FWHM, nm
1550	54	24	16	2,2
740	11	60	45	6,4

The losses in the waveguide circuit reduced the on-chip detection efficiency to 19%±6% and 21±14% for wavelengths of 1550 nm and 740 nm, respectively, as well as coupling losses restrict the system detection efficiency to the 2% @1550 nm and 0.42% @ 740 nm. By optimizing the waveguide circuit and using more efficient coupling with the optical fiber, the system detection efficiency can be significantly improved in the future.

This work was supported by Russian Science Foundation project 16-19-10633.

References

1. G. Gol'tsman, O. Okunev, G. Chulkova et al., Appl. Phys. Lett. **79**, 705 (2001)
2. C. M Natarajan, M. G Tanner, R. H. Hadfield, Supercond. Sci. Technol. **25**, 063001 (2012)
3. W. Pernice, C. Schuck, O. Minaeva et al., Nat. Commun. **3**, 1325 (2012)

Single photon detection in micron scale NbN and α-MoSi superconducting strips

Yuliya Korneeva[1,*], *Denis* Vodolazov[2], *Irina* Florya[1], *Nadegda* Manova[1], *Eugenii* Smirnov[1], *Alexander* Korneev[1,3], *Michail* Mikhailov[4], *Gregory* Goltsman[1,3] and *Teunis* M. Klapwijk[5,1]

[1]Moscow State Pedagogical University, 119991, Moscow, Russia
[2]Institute for Physics of Microstructures, Russian Academy of Sciences, 603950, Nizhny Novgorod, Russia
[3]Higher School of Economics National Research University, 101000, Moscow, Russia
[4]B. Verkin Institute for Low Temperature Physics and Engineering of the National Academy of Sciences of Ukraine, Kharkiv, Ukrane
[5]Kavli Institute of Nanoscience, Delft University of Technology, 2628 CJ Delft, Netherlands

Abstract. We experimentally demonstrate the single photon detection in straight micrometer-wide NbN and α-MoSi bridges. Width of the bridges is 2 μm, while the wavelength of the photon changes from 408 to 1550 nm and critical current exceeds 50% of the depairing current. Obtained results offer the alternative route for design of detectors without resonator and meander structure and indirectly confirm vortex assisted mechanism of single photon detection.

Until recently the operation of superconducting single photon detectors SSPD [1] was described by the "geometric hot spot model" first presented in Ref. [2] and improved further in subsequent works (a good review is given in Ref. [3]). In the framework of this model it is assumed that the single photon detection occurs only in a superconducting strip with a width comparable to the size of the hot spot, being in the range of 50 to 150 nm. It has led to devices consisting of fairly narrow and very long strips arranged in such a way that they fill a much larger area for good optical coupling. Recently, a more thorough theoretical approach [4] was applied to the problem of photon-detection in a superconductor. It is based on the standard theory of non-equilibrium superconductivity capable of treating processes depending on space and time. It predicts that in samples with a supercurrent up to about 0.7 to 0.8 of the depairing current, and uniform over the cross-section of the strip, the detection efficiency is not dependent on the width of the strip. We have experimentally demonstrated single-photon detection in 5-micron-wide constriction-type bridges made of polycrystalline NbN, using radiation of wavelength ranging from 408 nm to 1550 nm [5].

In this work we obtain the similar results for straight 2-μm-wide and 10-μm-long NbN strips. We measure close to unity internal detection efficiency (Fig. 1) at wavelengths 408 nm and 637 nm in agreement with the theoretical expectations [4]. As an additional check on the generality of this finding we also made wide strips of the amorphous silicide α-

* Corresponding author: korneeva@rplab.ru

MoSi. Unlike polycrystalline films, this amorphous material can be brought more easily in the required regime of a uniform superconducting gap and supercurrent. Additionally, amorphous α-MoSi is claimed to be attractive due to its lower critical temperature T_c compared to NbN which allows to obtain larger hot spot (for the same photon energy and film thickness). Experimentally α-MoSi wide bridge demonstrates IDE close to unity at wavelength up to 1 μm (Fig.1 b). We will present the results of the single-photon detection with the emphasis on the straight strips and compare them with the theoretical predictions.

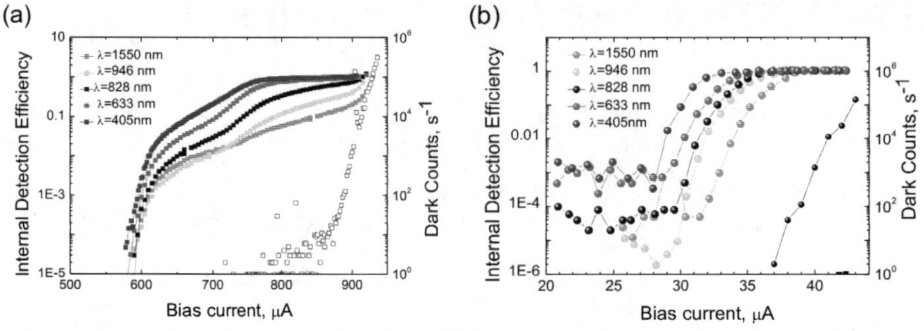

Fig. 1. Internal detection efficiency (detection efficiency normalized to absorption) and dark counts rate of NbN (a) and MoSi (b) samples.

Table 1. Parameters of the studied samples at the temperature *T*=1.7 K. The thickness of the film d was determined from time of deposition and deposition rate, w, width of the bridge, T_c is the critical temperature determined from the midpoint of the resistive transition, ρ(20K) is resistivity at *T*=20 K, j_c is critical current density, j_{dep} is the calculated depairing current density at *T*=1.7 K, using $j_{dep}(0)$ the calculated critical depairing current at $T = 0$ following from Eq.1 and Eq 2 from Ref.[5]

Sample	d (nm)	w (μm)	T_c (K)	ρ(20K) (μΩ cm)	D (cm^2 s^{-1})	j_c (A/cm^2)	j_{dep} (A/cm^2)
NbN	5.8	2	8.35	396	0.31	5.3×10^6	7.8×10^6
MoSi	2.8	1	3.1	203	0.19	0.79×10^6	1.7×10^6

The work is supported by the Russian Science Foundation (RSF) Project No.17-72-30036.

References

1. C. M. Natarajan, M. G. Tanner, and R. H. Hadfield, Supercond. Sci. Technol. **25,** 063001 (2012)
2. A. Semenov, G. Goltsman and A. Korneev, Physica C, **351**, 349 (2001)
3. A. Engel, J J Renema, K Il'in and A. Semenov Supercond. Sci. and Technol., **28,** 114003 (2015)
4. D. Y. Vodolazov, Phys. Rev. Appl. **7**, 034014 (2017)
5. Yu.P. Korneeva et. al. Phys. Rev. Appl. **9**, 064037 (2018)

EPJ Web of Conferences **190**, 04011 (2018) https://doi.org/10.1051/epjconf/201819004011
HBSM-2018

Cryogenic and large-base Fabry-Perot cavities for ultra-stable laser systems

Denis S. Kryuchkov[1,*], N.O. Zhadnov[1], K.S. Kudeyarov[1], I.A. Semerikov[1], K.Yu. Khabarova[1], and N.N. Kolachevskiy[1,2]

[1]P. N. Lebedev Physical Institute of the RAS, Moscow 119991, Russia
[2]Russian Quantum Center, Skolkovo, Moscow 143025, Russia

Nowadays most of high-precision spectroscopy experiments are based on lasers with spectral linewidth less than 1 Hz. Best systems now provide fractional frequency instability 4×10^{-17} on 1-100 seconds averaging time [1]. Such systems are a "hart" of most accurate frequency standards: their frequency is locked to ultra-narrow clock transitions in atomic and ionic ensembles. Stability of optical clock on short averaging times (less than ensemble preparation time) is completely defined by clock laser. On longer times Dick effect, that is determined by level of laser noise, limits stability as well.

Basic technique to obtain ultra-low fractional instability is locking laser frequency to the mode of a passive monolith ultra-stable high-finesse Fabry-Perot cavity. It has to be vibro-isolated and thermostabilized, is usually kept in ultra-high vacuum. Under those conditions fractional instability is fundamentally limited by thermal noise of cavity's mirrors (i. e. instability of length).

Allan variance of thermal noise — σ_y — can be used to predict best achievable fractional instability of cavity's eigenmode. In dependency on major parameters it is given as:

$$\sigma_y \propto \frac{\sqrt{T \times \phi_{coat}}}{L^{5/4} \times \lambda^{1/2} \times E^{1/2}}$$

Here, T — temperature, ϕ_{coat} — loss angle of mirror coating, L — cavity length, λ — laser wavelength, E — Young's modulus of mirror's substrate.

Our main goal is to improve efficiency of ultra-stable laser systems, and, as a result, to provide laser systems with fractional frequency instability about 10^{-16} on 1 second averaging time. According to equation for σ_y, main ideas to diminish thermal noise influence are: using high-Q materials, low temperatures and enlarging base of reference cavity. In present work we consider four ultra-stable systems.

The idea of lowering temperature with stabilization at the point of zero extension is presented by two cryogenic single-crystal silicon cavities – one with dielectric SiO_2/Ta_2O_5 mirrors (1) and one with crystalline GaAs/AlGaAs mirrors (2). Silicon is a very promising material for upcoming ultra-stable cavities because it has a temperature of zero extension at 124 K and has a very low mechanical loss angle. It is also transparent for 1.5 μm wavelength,

* Corresponding author: denis.kryuchkov@phystech.edu

which is covered by convenient erbium fiber lasers. Accurate calculations, proceeded for silica systems, resulted in following fundamental fractional instability limits: 3×10^{-16} and 6×10^{-17} for cavities with dielectric (1) and crystalline mirrors (2) respectively [2]. We have systems (1,2) assembled at this moment. (1) has finesse about 580'000, measured by ring-down method; (3) — about 250'000. Both silica cavities are mounted in UHV cryogenic chambers and stabilized at 124K with precision better than 3 mK. Fractional instability characterization via frequency correlation for two silica cavities will be performed in future.

The idea of enlarging cavity's base is presented by two large vertical and horizontal (3,4) cavities made of Ultra-low Expansion glass, operating on 698 nm wavelength. These systems are being assembles at this moment. Fundamental frequency instability limit calculations resulted in 7×10^{-17} for both large systems [2]. Also, vertical and horizontal cavities were numerically analyzed with finite element method to find out proper shape and optimal vibration-immune suspension configuration. Was found out that biconical is the best shape for (3): it affords least vibration susceptibility. For system (4) vibration-immune suspension system was developed, where bearings are located at the Airy points. Deformations under the action of horizontal and vertical accelerations were simulated for (3,4) as well.

Presented systems (3,4) will be used as clock lasers in Sr optical lattice clock in VNIIFTRI [3], systems (1,2) — as optical frequency synthesizer stabilizers and for stabilizing long fiber-optic communication lines.

References

1. D. Matei et al., Phys. Rev. Lett. **118**, 263202 (2017)
2. N. Zhadnov et al., Quantum Electronics, **48**, 425 (2018)
3. O. Berdasov et al., Quantum Electronics, **47**, 400 (2017)

EPJ Web of Conferences **190**, 04012 (2018) https://doi.org/10.1051/epjconf/201819004012
HBSM-2018

Spontaneous transitions to enhanced fluorescence for GeV centers in a single microcrystalline diamond

Natalia A. Lozing[1,2,*], *Maxim G.* Gladush[1,3], *Ivan Yu.* Eremchev[1], *Eugeniy A.* Ekimov[4], and *Andrey V.* Naumov

[1]Institute for Spectroscopy RAS, 5 Fizicheskaya Str., Troitsk, Moscow 108840, Russia
[2]National Research University Higher School of Economics, 20 Myasnitskaya Str., Moscow 101000 Russia
[3]Moscow State Pedagogical University,1/1 M. Pirogovskaya Str., Moscow 119991, Russia
[4]Vereshchagin Institute for High Pressure Physics of the Russian Academy of Sciences, 14 Kalushskoye Shosse, Troitsk, Moscow 108840, Russia

Abstract. We propose a theoretical explanation of spontaneous transitions between dim and bright fluorescence intensity states observed experimentally in a microcrystal of diamond with germanium-vacancy colour centres driven by a continuous wave laser. We use a generalized system of optical Maxwell-Bloch equations derived for an emitter in an ensemble of motionless similar particles embedded in a dielectric medium, which is transparent for the incident light. A numerical analysis of transient regimes and several models of slow damping of the bright luminescence mode are reported.

Spontaneous flickering of the fluorescence intensity was registered experimentally for a microcrystal of diamond with germanium-vacancy centres excited by a continuous wave laser light. The fluorescence intensity from the sample was observed to be switching to several bright modes and returning to the only well determined dim mode at random occasions on the time-scale of seconds. The effect demonstrated smooth transition dynamics on the same scale and tended to last for hours. The phenomenon had a strict power threshold. These characteristics are far from the effect of "blinking" but are similar to experiments in micron-sized crystals of methylammonium lead bromide and CaS:Eu2+ crystals [1-3]. It can be assumed that the observed effect is likely to be due to nonlinear interactions of the incident light with the photoactive sites inside the crystal. It is known that the optical properties of dense atomic ensembles or complex materials can differ greatly from the optical properties exhibited by independent quantum emitters. In this report we propose a theoretical explanation based on the possibility for a cooperative ensemble of emitting centres to show intrinsic cooperative optical multistability [4]. It is shown that the Maxwell-Bloch type equations describing such a system [5], even in the presence of large values of the phase relaxation rate, retain the possibility of having several stable steady state solutions that form a multistable optical response for certain

* Corresponding author: lozing@isan.troitsk.ru

combinations of the pump power and concentration of the emitters. A numerical analysis has shown that the transition time may exceed the lifetime by an order of magnitude. We report several models of slow damping of the bright fluorescence mode and discuss alternative reasons for such a behaviour.

This work was supported by the grant RFBR 16-02-01174.

References

1. S. P. Feofilov and A. B. Kulinkin, J. Lumin., **170**, 121 (2016)
2. A. Halder, N. Pathoor, A. Chowdhury, and S. K. Sarkar, J. Phys. Chem. C, **122**, 15133 (2018)
3. N. Pathoor, A. Halder, A. Mukherjee, J. Mahato, S. K. Sarkar, and A. Chowdhury, Angew. Chemie Int. Ed., **57**, 11603 (2018)
4. F. A. Hopf, C. M. Bowden, and W. H. Louisell, Phys. Rev. A, **29**, 2591 (1984)
5. M. G. Gladush, D. V. Kuznetsov, and V. K. Roerich, Eur. Phys. J. D, **64**, 511 (2011)

EPJ Web of Conferences **190**, 04013 (2018) https://doi.org/10.1051/epjconf/201819004013
HBSM-2018

Inhomogeneous broadening and spectral diffusion of the CdSe nanocrystals in the solid nanocomposites

Konstantin Magaryan[1,2,*]

[1]Moscow State Pedagogical University, 119991 Moscow, Russia
[2]Institute for Spectroscopy RAS, 108840, Moscow, Russia

Abstract. The mechanisms of broadening of optical spectra of CdSe nanocrystals (quantum dots, QD) synthesized in the liquid crystalline matrix were studied. The experiment confirms that the luminescence spectra of macroscopic samples are broadened inhomogeneously. In addition, small fractions of nanocomposites (with single QDs and small ensembles of QDs) manifest noticeable stochastic spectral diffusion.

The object under study is nanocomposite with CdSe nanocrystals (quantum dots, QD). Nanocrystals were grown in the ionic liquid crystalline matrix of cadmium alkanoate using template synthesis [1]. It's quite a new method of synthesis QDs inside liquid crystalline host medium. The resulting object combines peculiarities of semiconductor heterostructures and the liquid crystal anisotropic medium. In emission spectra of ensemble [2, 3] samples under study show two separated bands in vicinity of exciton transition and wide band in the long wavelength region (see Fig. 1*a*). In this talk we report on luminescence spectra of small fractions of such nanocomposites, which include single QDs and/or small ensembles of QDs, as well as about comparative analysis with the spectra of macroscopic samples.

Site-selective spectroscopy is a very powerful tool for detailed study of dynamics in a doped disordered solids [4, 5]. The luminescence and absorption (fluorescence excitation) spectra for the samples were obtained using luminescence microscopy technique and reported in [2, 3]. The luminescence spectra of small fractions of the sample (isolated micro- and nanoclusters which contains single QD or their small ensembles) were obtained with the help of setup for a spatially resolved single molecule detection system based on the fluorescent microscope [4, 6] with the home-built prism spectrometer. The samples were prepared by spin-coating of low concentrated colloidal toluene solutions with nanocomposites powder. The luminescence was excited by 405 nm cw-laser. Luminescence signal was collected in confocal scheme from isolated nano-/micro-particles.

The luminescence spectra from different particles (Fig. 1*b*) demonstrate strong inhomogeneous broadening of the macroscopic sample. On the figure 1 (*c*) several examples of the recorded photoluminescence spectra of the same diffraction-limited area (some selected isolated particle) are shown. Effect is observed under continuous cw-laser excitation at room temperature. Such random jumps in the emission wavelength called spectral diffusion lead to the broadening of the nanocrystals ensemble luminescence

* Corresponding author: ka.magaryan@mpgu.edu

spectra. Luminescence from QDs with different size going along with phenomenon of spectral diffusion leads to inhomogeneous broadening of the spectra of the composite.

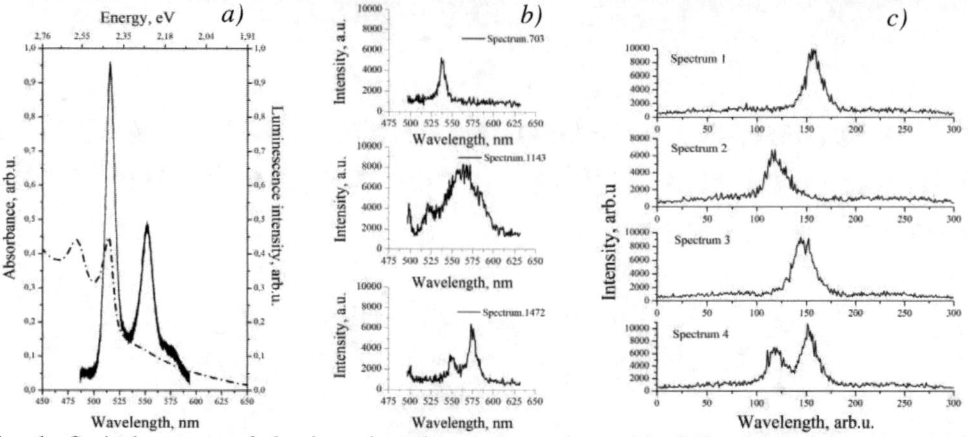

Fig. 1. Optical spectra of the investigated nanocomposites with CdSe QDs (2.3 nm). (*a*) The macroscopic sample. (*b*) Three different isolated small nm/µm-sized fractions of the sample. (*c*) The same nm-sized particle (with near-single QD) as measured sequentially in different time moments.

There is a possibility to analyze relation of such spectral diffusion process to single QD blinking process [7-10]. In fact, the fluorescence trace for single QDs can be build. We observe the random duration of the on\off-intervals. When several fluorescence tracks are summarized, the effect of fluorescence blinking is no longer observed. Events fluctuations are complex with different amplitudes. If we propose that there are several emitting particles, then the probability of their simultaneous switch to the off-state would be low.

The photon distribution function for a single QD can be reconstructed from a fluorescence trace. As we have found for many cases the complicated fluorescence intermittency pattern cannot be explained in the context of the Efros-Rosen charge model [7]. In our case long on-intervals are absent and the fluorescence fluctuates between multiple states with different intensity. We describe such blinking in the general framework of the two models: multiple recombination center model [8] or combined model based on the modified charging model and the two-level system model [9,10]. This dynamics can be described using second approach that takes into account ionization/neutralization processes in a nanocrystal and fluctuations of atoms on a QD interface.

This work was supported by RFBR (proj. no 18-02-01121 – study of spectral diffusion in a new luminescent nanomaterials) and Scientific Council of the President of the Russian Federation (proj. no. MK-342.2017.2 – study of the broadening of the nanocomposites spectra). The data for analysis was obtained in collaboration with Dr. I.Yu. Eremchev and Prof. A.V. Naumov. The author is grateful to them as well as to Prof. I.A. Vasilieva for the fruitful discussions.

References

1. D. Zhulai et al., NRL, **10**, 66 (2015)
2. K.A. Magaryan et al., J. Lumin., **169**, 799 (2016)
3. K.A. Magarian et al., J. Phys: Conf. Ser., **478**, 012007 (2013)
4. A.V. Naumov, Physics Uspekhi, **56**, 605 (2013)
5. K.R. Karimullin, A.V. Naumov, J. Luminescence, 152, 15 (2014)
6. A.V. Naumov, I.Yu. Eremchev, A.A. Gorshelev, Eur. Phys. J. D, **68**, 348 (2014)
7. A.L. Efros, M. Rosen, Phys. Rev. Lett., **78**, 1110 (1997)
8. P.A. Frantsuzov, S. Volkan-Kacso, B. Janko, Phys. Rev. Lett., **103**, 207402 (2009)
9. I.S. Osad'ko, I.Y. Eremchev, A.V. Naumov, J. Phys. Chem. C, **119**, 22646 (2015)
10. I.Y. Eremchev, I.S. Osad'ko, A.V. Naumov, J. Phys. Chem. C, **120**, 22004 (2016)

EPJ Web of Conferences **190**, 04014 (2018) https://doi.org/10.1051/epjconf/201819004014
HBSM-2018

Switching or blinking? – The switching behaviour of single photochromic triads

Johannes Maier[1,*], *Martti* Pärs[1], *Tina* Weller[2], *Mukundan* Thelakkat[2], and *Jürgen* Köhler[1]

[1]Spectroscopy of soft Matter, University of Bayreuth, 95447 Bayreuth, Germany
[2]Applied Functional Polymers, University of Bayreuth, 95447 Bayreuth, Germany

Abstract. Photochromic molecules can be interconverted between two bistable conformations by light [1–3]. Irie and coworkers described a strategy to achieve superior fluorescence characteristics and outstanding switching characteristics of a photochromic unit by linking strong fluorophores covalently to photochromic building blocks [3,4]. Accordingly, we synthesised molecular triads that consist of two perylene bisimide (PBI) fluorophores covalently linked to a dithienylcyclopentene (DCP) photochromic switch, see fig. 1. Such kinds of triads are promising candidates for super-resolution microscopy like RESOLFT and PALM [5,6], or can be used as optical transistors or memories [4,7].

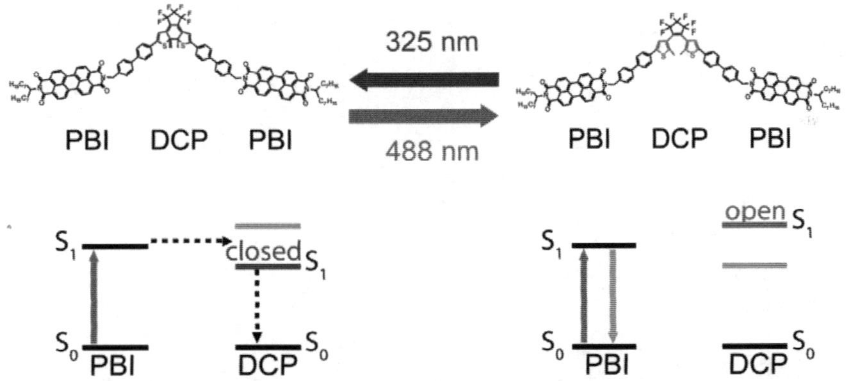

Fig. 1. Top: closed (left) and open (right) state of the photoswitchable triad PBI-DCP-PBI, where the closed state is reached via illumination in the UV spectral range (325 nm) and the open state is reached via illumination in the visible range (488 nm). Bottom: Energy level scheme of the closed (left) and open (right) state of the triad. In the closed state, the excited state of the switch DCP lies below the excited state of the PBI, energy transfer is possible and the fluorescence of the PBI is reduced, whereas in the open state of the switch, energy transfer is not possible and the PBI shows fluorescence.

Regarding super-resolution microscopy and a demand for miniaturization for transistors and memories, single molecule applications are of great interest [8]. When studying the

* Corresponding author: joh.maier@uni-bayreuth.de

fluorescence of single photochromic triads under simultaneous illumination of visible and UV light, the on/off-durations of the telegraph-like fluorescence intensity vary as a function of the UV intensity [9]. The problem faced is that it is impossible to discriminate between off-states that are due to illumination with UV light and those that are due to spontaneous, yet unwanted and unavoidable blinking processes. However, a statistical analysis allows to determine the probability for transitions to the off state by deliberate switching of the triad. Such an analysis yields the positive predictive value $p(S|+)$, that corresponds to the conditional probability that the triad indeed has been switched (S) given that a delay (+) between the onset of the excitation laser and the onset of the fluorescence has been observed. For the 15 triads we examined, this probability amounts on average to 0.8 ± 0.1 with a maximum of approximately 0.96 [10], sufficiently large for application in super-resolution microscopy.

References

1. H. Bouas-Laurent, H. Dürr, Pure and Applied Chemistry, **73**, 639 (2001)
2. B.L. Feringa, *Molecular Switches* (Wiley-VCH, Weinheim, 2001)
3. M. Irie, T. Fukaminato, K. Matsuda, S. Kobatake, Chem. Rev., **114**, 12174 (2014)
4. M. Irie, Chem. Rev., **100**, 1685 (2000)
5. O. Nevskyi, D. Sysoiev, J. Dreier, S.C. Stein, A. Oppermann, F. Lemken T. Janke, J. Enderlein, I. Testa, T. Huhn, D. Wöll, Small, **14**, 1703333 (2018)
6. M. Irie, M. Morimoto, BCSJ, **91**, 237 (2018)
7. M. Pärs, C.C. Hofmann, K. Willinger, P. Bauer, M. Thelakkat, J. Köhler, Angew. Chem. Int. Ed., **50**, 11405 (2011)
8. M. Irie, T. Fukaminato, T. Sasaki, N. Tamai, T. Kawai, Nature, **420**, 759 (2002)
9. T. Fukaminato, T. Umemoto, Y. Iwata, S. Yokojima, M. Yoneyama, S. Nakamura, M. Irie, J. Am. Chem. Soc., **129**, 5932 (2007)
10. J. Maier, M. Pärs, T. Weller, M. Thelakkat, J. Köhler, Sci. Rep., **7**, 41739 (2017)

Control of the direction of energy transfer in associates of colloidal quantum dots Ag_2S/TGA and dye molecules

Oleg V. Ovchinnikov*, *Mikhail S.* Smirnov, *Irina G.* Grevtseva, *Tamara S.* Kondratenko and *Aleksey S.* Perepelitsa

Voronezh State University, Department of optics and spectroscopy, 394018 Voronezh, Russia

Abstract. The possibilities for changing the direction of energy transfer in hybrid associates of colloidal Ag_2S quantum dots (QDs) and thionine molecules are analyzed. The studies were performed by transmission electron microscopy, absorption and luminescence spectroscopy, PL decay study (time correlated single photon counting). An increasing of the average QDs size from 1.8 nm to 5.5 nm, and also a shift of the luminescence band from 630 nm to 950 nm, were found at using of TGA molecules and sodium sulfide as a sulfur precursor. Hybrid association of QDs (1.8 nm) with TH^+ molecules leads to quenching of QDs luminescence with a simultaneous reduction of the luminescence lifetime from 13.7 to 6.5 ns. An association of QDs with a luminescence band maximum of 950 nm with TH^+ molecules leads to quenching of TH^+ luminescence and a reduction in its lifetime of luminescence from 0.43 to 0.3 ns. It was concluded that the reduction of lifetime of luminescence caused by the resonant nonradiative energy transfer between the components of the associates. An increasing in the average size of QDs leads to a change in the direction of energy transfer between the components of the associates.

In recent years, organo-inorganic associates of colloidal semiconductor quantum dots (QDs) and dye molecules (Dye) have been actively studied. They are used as luminescent sensors, photosensitizers of singlet oxygen, disinfecting systems. This is due to the appearance of new properties for associates that are not characteristic of individual components. The main role in the formation of new properties is played by the processes of exchange of electronic excitations.

Currently, the most detailed cases are considered when energy donors are QDs. Dye molecules are acceptors. The main energy transmission channels are: nonradiative Forster resonance energy transfer (FRET), transport of charge carriers. Herewith it has been showed the principal possibility of nonradiative energy transfer from QDs excitons to the dye and from the center of recombination luminescence to the dye. The situation when dyes are donors of an energy, and QDs are acceptors, is practically not considered.

* Corresponding author: Ovchinnikov_O_V@rambler.ru

The results of the study of nonradiative energy transfer in hybrid associates of colloidal Ag_2S quantum dots, passivated with thioglycolic acid (Ag_2S/TGA) and thionine molecules (TH^+) are presented below.

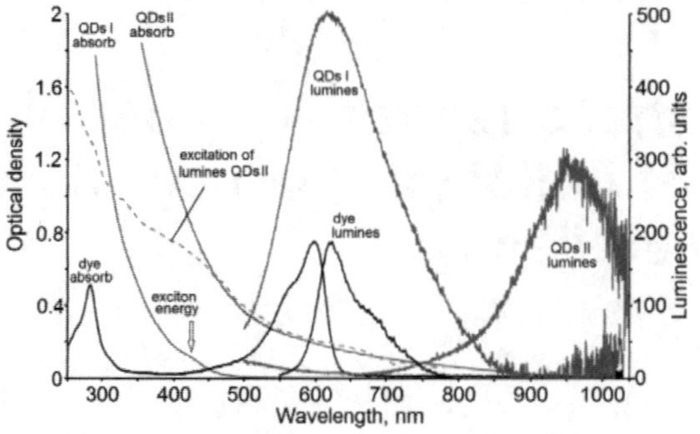

Fig. 1. Absorption and luminescence spectra of the components of hybrid associates.

Figure 1 shows the absorption and luminescence spectra of the components of hybrid associates. Samples of QDs of I and II type have luminescence bands with maxima located in the region of 630 nm and 950 nm, respectively. For type I QDs, the luminescence band overlaps with the absorption band of thionine. For type II QDs, the luminescence band is located on the long-wave side of the absorption and luminescence bands of thionine.

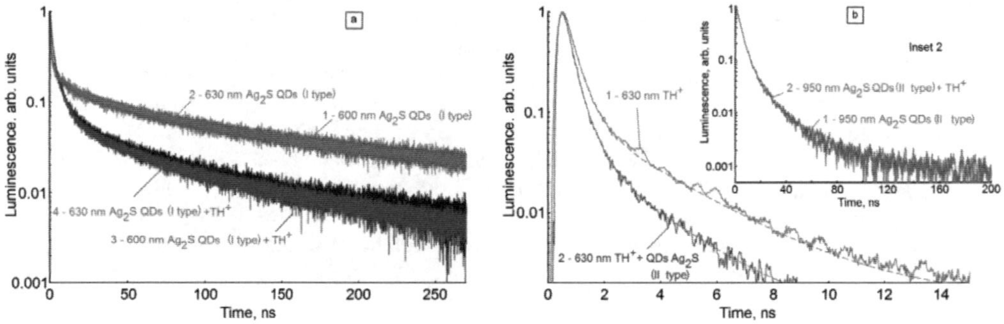

Fig. 2. Decay of luminescence curve.

At assembling of hybrid associates noticeable changes occurred in the absorption and luminescence properties of the components. The absorption spectra of thionine molecules increases in full width at half maximum toward shorter wavelengths to a few nanometers. For associates of type I QDs, thionine molecules were assembled into H aggregates. Also, the intensity and lifetime of the QDs luminescence decreased and simultaneously the luminescence of thionine increased. For associates of type II QDs, luminescence of thionine was quenched and luminescence of QDs increased. In addition, the lifetime of the luminescence of thionine molecules was reduced.

Thus, the quenching of luminescence and the shortening of its lifetime for one component and the increasing of luminescence of another component unambiguously indicate a nonradiative transfer of energy. In this case, the direction of the transfer is different for associates from the I and II type QDs.

This work was supported by the grant of the RFBR (proj. no. 18-52-00037 Bel_a).

Nonlinear optical absorption of non-spherical silver nanoparticles and organic dyes mixtures

Andrey I. Zvyagin, *Aleksey S.* Perepelitsa, *Mariya S.* Lavlinskaya, *Oleg V.* Ovchinnikov*, *Mikhail S.* Smirnov, and *Rashid A.* Ganeev

Voronezh State University, Department of optics and spectroscopy, 394018 Voronezh, Russia

Abstract. We study the features of the nonlinear optical response of silver nanoparticles (Ag NPs) of spherical and triangular shapes synthesized using a citrate-borohydride approach and their mixtures with methylene blue (MB^+). The Z-scans at the wavelengths of 532 nm using 10 ns pulses were used to analyze the nonlinear optical absorption of MB^+ solutions, Ag NPs of different morphology, particularly, the concurrence of saturable and reverse saturable absorption in the solution of triangular Ag NP and MB^+. We show the significant contribution to the increasing in the normalized transmittance accomplished by photochemical degradation of the non-spherical Ag NPs (extinction maximum of 600 nm) and transformation into the spherical ones (extinction maximum of 400 nm). The nonlinear light scattering is also analyzed, which showed the contribution to the variation of the sample transmission when it approaches the focal plane. It is shown that the nonlinear optical response is not additive in the mixture of Ag NPs and MB^+. The increasing of the nonlinear absorption is probably due to both a change in the population of the triplet states of the dye and the efficiency of photochemical degradation of NPs in the presence of MB^+.

At present, research in the field devoted to the development of control techniques for nonlinear optical properties of organo-inorganic nanostructures is relevant. The occurrence of interaction manifestations in other properties, in particular nonlinear optical, is of undoubted interest, including a practical one. For mixtures of plasmon nanoparticles (NPs) and dyes, studies of this kind are virtually unknown in the literature. In particular, the presence of plasmon NPs near MB^+ can affect the probabilities of optical transitions and the mechanisms of photophysical processes in the molecule. However, the effect of plasmonic NPs on the two-photon absorption (TPA), the saturable absorption (SA), and the reverse saturable absorption (RSA) of MB^+ has not been performed yet. In this paper, we report the studies of the nonlinear absorptive characteristics of mixtures of Ag NPs and MB^+ in colloidal solutions using 10 ns pulses at the wavelength of 532 nm.

The measurements of the nonlinear optical properties of our samples were carried out by the standard Z-scan technique. The Nd:YAG laser generated single pulses (λ = 1064 nm, τ=10 ns) at 1 Hz repetition rate. The probe second harmonic (λ=532 nm) of this radiation generated in KDP crystal was focused by a 300-mm focal length spherical lens. The laser pulse energy was measured by a first calibrated photodiode and then registered by a digital

* Corresponding author: Ovchinnikov_O_V@rambler.ru

voltmeter. The 5-mm-thick glass cells containing studied samples were moved along the z-axis through the focal plane of probe pulse using a translating stage. The open-aperture Z-scan scheme, i.e. the one without aperture in front of second photodiode allowed measuring all transmitted radiation to determine the dependence of the normalized transmittance on the position of sample on the Z-axis. To observe the contribution to the Z-scan from nonlinear dynamic scattering, a scheme was also used in which a second diode was installed at an angle of 4.5° with regard to the optical axis.

We used Ag NPs obtained during the synthesis process, in which the reducing agents were trisodium citrate and sodium borohydride. According to poly-N-vinylpyrrolidone and sodium citrate act as stabilizers preventing agglomeration of NPs during synthesis. The use of several stabilizers promotes the growth of non-spherical Ag NPs due to their ability to limit the growth of certain crystallographic planes of Ag NPs. Hydrogen peroxide, in turn, provided the necessary assistance to induce further oxidation of small particles to form NPs of different sizes and shapes.

The size and geometry of the synthesized ensembles of colloidal Ag NPs were determined using a transmission electron microscope (Libra 120, CarlZeiss). Size distribution for the ensemble of Ag NPs was obtained by digital analysis of TEM images. The extinction spectra of Ag NPs of different morphology were studied using the spectrophotometer USB2000+ with a source of radiation USB-DT (Ocean Optics).

It is shown that it is possible to obtain ensembles of spherical Ag NPs with average dimensions of 9±4 nm, whose extinction maximum was located near 420 nm. The NPs concentration in the solution was 4×10^{14} cm^{-3}. Further holding of the colloidal solution for 30 minutes led to a change in the form of distribution of NPs and the appearance of two maxima on the histogram at 7 and 17 nm. The extinction maximum for such particles was located near 520 nm. The geometry of the NPs became non-spherical, resembling elongated droplets. The NPs do not have time to become triangular. Further holding of the colloidal solution for 60 minutes led to a change in size to 15±10 nm and a predominantly triangular geometry. The extinction maximum of these NPs was located near 600 nm.

We have demonstrated the dynamics of the nonlinear optical response of Ag NPs of spherical, mixed and triangular forms. It is shown that the reduction of the normalized transmittance in the focal plane of Z-scan can be induced, alongside the RSA, by photochemical degradation of the non-spherical NPs characterized by the plasmon resonance near 600 nm. Their transformation into spherical NPs characterized by the plasmon resonance near 420 nm can lead to a decrease in the optical density at a wavelength of 532 nm. We also analyzed the dynamic scattering of laser radiation, which may contribute to the formation of a symmetric dip in the Z-scan when samples with NPs approach the focal plane.

In mixtures of Ag NPs and MB^+ in colloidal solutions using 10 ns pulses at the wavelength of 532 nm the enhancement of nonlinear response was observed. It was manifested in the enhancement of symmetric dip in the Z-scan which observed when sample approaches the focal plane. In this case, the intrinsic nonlinear response from the analogous volume of the dye without NPs was practically absent. The observed feature can be explained by a number of processes. On the one hand, near Ag NPs the changing in the probabilities of optical transitions of MB^+ molecules is probably. In this case, such an action is probably not only for singlet transitions, but also for triplet transitions. The change in the population of the triplet states can affect the enhancement of the RSA in MB^+. On the other hand, the optical absorption spectra indicate a substantial rearrangement of the mixtures. Probably photochemical transformation occurs both the dye and NPs, especially since for NPs in the absence of MB^+, the transformation of their geometry is observed.

Results obtained in the course of the state task of the Ministry of Education and Science of the Russian Federation in the field of research activity 2017-2019 (Project №3.6672.2017/8.9).

Spectral properties of hybrid associates of colloidal quantum dots $Zn_{0.5}Cd_{0.5}S$, europium tenoyltrifluoroacetonate and methylene blue

Mikhail S. Smirnov*, *Aleksey S.* Perepelitsa, *Tamara S.* Kondratenko, *Andrey I.* Zvyagin and *Oleg V.* Ovchinnikov

Voronezh State University, Department of optics and spectroscopy, 394018 Voronezh, Russia

Abstract. Hybrid associates formed from colloidal $Zn_{0.5}Cd_{0.5}S$ quantum dots, passivated with thioglycolic acid, europium tenoyltrioteracetonate and methylene blue molecules, absorption, luminescence, IR and time-resolved spectroscopy technique are studied. The shift of the IR absorption bands of COO^- and C=O groups of thioglycolic acid and europium thenoyltrifluoroacetonate has been detected. An increase in the efficiency of excitation of europium luminescence and a simultaneous increase lifetime of its luminescence upon adsorption on $Zn_{0.5}Cd_{0.5}S$ quantum dots were found. Addition of methylene blue (thionine) molecules leads to quenching of the trap state luminescence of $Zn_{0.5}Cd_{0.5}S$ and intracentric luminescence of Eu^{3+}. A conclusion about the adsorption of Eu^{3+} on the surface of $Zn_{0.5}Cd_{0.5}S$ quantum dots and the nonradiative energy transfer to methylene blue molecules was made.

Semiconductor crystals and dielectrics, doped with rare earth element (RRE), find extensive applications as materials for solid-state lasers, fiber amplifiers, biolables, solar cells, etc. Quantum dots (QDs) doped with RRE, whose own optical and electronic properties have a size dependence, have considerable interest. Among them are QDs doped with europium ions. An additional modification of the optical properties can be achieved by conjugating of doped quantum dots with molecules of organic dyes.

The fig. 1 shows the luminescence spectra of QDs of $Zn_{0.5}Cd_{0.5}S$/TGA in ethanol solution, and also their mixtures with europium tetrafluoroacetonate (Eu^{3+}:TTA), introduced at the stage of QDs crystallization. The luminescence in band with a maximum at 520 nm occurs with participation of trap states. In the luminescence spectrum of Eu^{3+}: TTA the peaks are present: at 592 nm, 615 nm, 653 nm and 702 nm - transitions between terms $^5D_0 \rightarrow {}^7F_1$, $^5D_0 \rightarrow {}^7F_2$, $^5D_0 \rightarrow {}^7F_3$ and $^5D_0 \rightarrow {}^7F_4$, respectively.

At introducing of Eu^{3+}:TTA into a colloidal solution with emerging QDs, the changes in luminescence properties were observed: - with increasing of QDs concentration, the luminescence intensity of Eu^{3+} in all bands initially increases by 5 times, and then decreases; - with increasing of concentration of Eu^{3+}:TTA, the intensity of trap state luminescence decreases strongly, and the luminescence lifetime decreases; - at increasing

* Corresponding author: Smirnov_M_S@mail.ru

of QDs concentration to 20%, luminescence decay in the Eu^{3+} band slows down; - the intensity of the emission Eu^{3+} grows at the initial point of decay of its luminescence, i.e. the effectiveness of its excitation is growing.

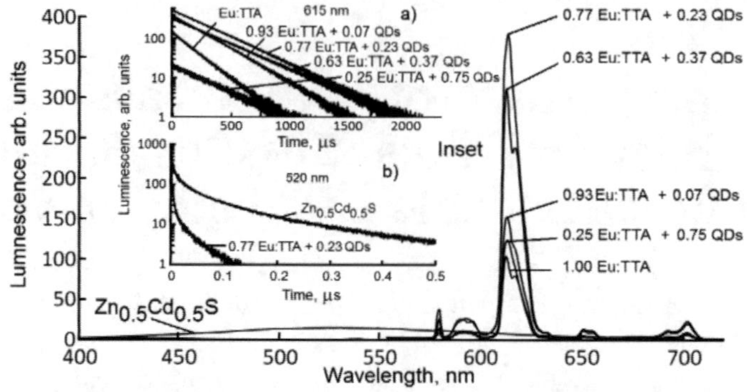

Fig. 1. Spectra of luminescence of $Zn_{0.5}Cd_{0.5}S$/TGA+Eu^{3+}:TTA. In the insert is decay of luminescence of $Zn_{0.5}Cd_{0.5}S$/TGA and Eu^{3+}:TTA.

It can be concluded that an association of QDs and Eu^{3+}: TTA is observed. Two types of similar associates are possible: i - some TTA molecules are replaced by TGA molecules and a carboxylate complex is formed. The structure of such a complex can be represented in the form $Zn_{0.5}Cd_{0.5}S$/TGA/Eu^{3+}. ii - Eu^{3+} is adsorbed to QDs and integrated into the near-surface layer of QDs. The structure of $Zn_{0.5}Cd_{0.5}S$/$Zn_{0.5}Cd_{0.5}S$:Eu^{3+}:TTA (TGA), passivated by TGA and TTA molecules, is formed.

According to the IR absorption spectra, it was established on the basis of insignificant shifts in the absorption bands of the carboxyl group of TGA and the carbonyl group of TTA that the formation of $Zn_{0.5}Cd_{0.5}S$/$Zn_{0.5}Cd_{0.5}S$: Eu^{3+}:TTA (TGA), i.e. doping of QDs with Eu^{3+} ions.

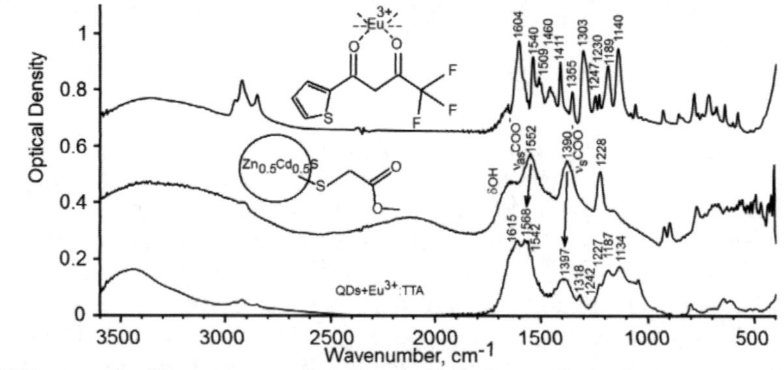

Fig. 2. FTIR spectra and structures of the investigated samples.

The conjugation of QDs of $Zn_{0.5}Cd_{0.5}S$/$Zn_{0.5}Cd_{0.5}S$:Eu^{3+}:TTA (TGA), with molecules of methylene blue (thionine) leads to quenching of trap state luminescence and intracentric luminescence and rise up of luminescence of the dye. At the same time, the luminescence lifetime in the bands of trap state luminescence and intracentric luminescence is reduced. A conclusion about the transfer of energy from the center of recombination luminescence and from the intracentral luminescence of Eu^{3+} is drawn.

The work was supported by the grant of the RFBR (No. 17-02-00748a).

https://doi.org/10.1051/epjconf/201819004018

Near-infrared photopolymerization assisted by upconversion nanophosphors for biomedical applications

Alexander G. Savelyev[1,2*], *Vladimir* A. Semchishen[1], *Andrey* V. Nechaev[1,3], *Kirill* V. Khaydukov[1], *Polina* A. Demina[1,4], *Alla* N. Generalova[1,4], and *Evgeny* V. Khaydukov[1,2]

[1]Federal Scientific Research Centre "Crystallography and Photonics" of the Russian Academy of Sciences, 119333 Moscow, Russia
[2]Sechenov First Moscow State Medical University, 119991 Moscow, Russia
[3]Institute of Fine Chemical Technologies, Moscow Technological University, 119571 Moscow, Russia
[4]Shemyakin-Ovchinnikov Institute of Bioorganic Chemistry of Russian Academy of Sciences, 117197 Moscow, Russia

Abstract. We present the concept and the experimental demonstration of near-infrared photopolymerization assisted by specially designed upconversion nanophosphors. The principle of this technique is based on conversion of 980 nm laser irradiation to ultraviolet photons subsequently absorbed by photoinitiator. The nonlinearity of upconversion allows for activation of the process locally in the laser beam waist. This approach enables precise fabrication of 3D constructs directly in the volume of photocurable composition. Furthermore, the presented technique is suitable for polymerization of a wide range of photocurable resins as well as gelation of hydrogels for biomedical applications.

A broad range of techniques based on photo-induced polymerization have been developed. The overwhelming majority of those employs the single-photon absorption. In this case, the polymerization of the exposed photocurable compositions (PCCs) is a linear process getting launched in every point within the irradiation penetration depth and depending only on exposure dose. This approach is widely used in laser stereolithography [1] and direct light processing [2] when the photocurable resin is patterned and recoated layer by layer. However, this technique is not always suitable for *in situ* formation of structures. In this case the two-photon photopolymerization can be implemented. The method is based on two-photon absorption of light by photoinitiator [3]. The nonlinearity of absorption process allows activation of polymerization locally in the laser beam waist. Furthermore, implementation of near-infrared light instead of UV exposure makes possible to increase the irradiation penetration depth. Two-photon polymerization represents a high resolution method for drawing structures directly in the volume of PCC. Nevertheless, the need to use complicated femtosecond laser systems in order to achieve the required high intensity in the beam waist limits the wide implementation of this technique.

* Corresponding author: a.g.savelyev@gmail.com

In this paper, we report on formation of structures in the volume of PCC using near-infrared photopolymerization. We propose to impregnate the photosensitive resin with upconversion core/shell $NaYF_4:Yb^{3+},Tm^{3+}/NaYF_4$ nanophosphors capable to absorb light at 980 nm and effectively convert it to ultraviolet lines (Fig 1 (left)) within the absorption band of photoinitiator launching the polymerization. The conversion coefficient was measured as $\eta_{UC}^{(UV)}=2\%$. In comparison to two-photon polymerization the excitation action is performed via real energy states. Therefore, formation of 3D structures in the depth of PCC volume can be carried out at moderate intensities using simple semiconductor lasers instead of expensive femtosecond systems.

Figure 1 (right) illustrates the scaffolds produced by near-infrared photopolymerization in the depth of PCC on the base of oligocarbonate methacrylate (OCM-2) containing 1 % of TPO photoinitiator and upconversion nanoparticles. The standard Irgacure or chlorine P6 can be also used as photoinitiators.

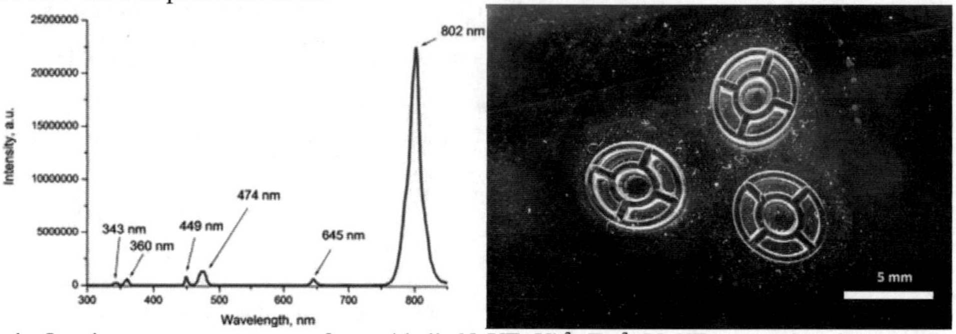

Fig. 1. Luminescence spectrum of core/shell $NaYF_4:Yb^{3+},Tm^{3+}/NaYF_4$ nanophosphors under excitation at 980 nm (left). Scaffolds on the base of oligocarbonate methacrylate produced by near-infrared photopolymerization (right).

Implementation of near-infrared photopolymerization assisted with upconversion nanophosphors has a great potential for gelation of biocompatible hydrogels, e.g. on the base of polyethylene glycol diacrylate. For this aim $NaYF_4:Yb^{3+},Tm^{3+}/NaYF_4$ nanophosphors can be modified with amphiphilic polymer in order to provide the hydrophilic properties. In this case the photoinitiator also should be water soluble. Therefore, the endogenous flavin mononucleotide [4] or chlorine P6 conjugated with polyethylene glycol can be the grate candidates. The ability to form the structures in the depth of biocompatible water-based PCCs makes the near-infrared photopolymerization assisted with upconversion nanophosphors extremely interesting for biomedical applications, when the polymerization process can be performed *in situ*.

This scientific work has been partly supported by Grants RFBR No 17-00-00118 in the part of synthesis of nanophosphors and RSF No. 17-19-01416 in the part of scaffolding.

References

1. F.P.W. Melchels, J. Feijen, D.W. Grijpma, Biomaterials **31**, 6121 (2010)
2. J.-W. Choi, R. Wicker, S.-H. Lee, K.-H. Choi, C.-S. Ha, I. Chung, J. Mater. Process. Tech. **209**, 5494 (2009)
3. O. Kufelt, A. El-Tamer, C. Sehring, S. Schlie-Wolter, B.N. Chichkov. Biomacromolecules **15**, 650 (2014)
4. A.G. Savelyev, K.N. Bardakova, E.V. Khaydukov, A.N. Generalova, V.K. Popov, B.N. Chichkov, and V.A. Semchishen, J. Photochem. Photobiol. A: Chemistry **341**, 108 (2017)

Structural and time-domain peculiarities of the fluorescence excitation spectra of single Mg-TAP in a polymer at low temperatures

Aleksandr Savostianov[1,*], *Ivan* Eremchev[1,2], *Alexei* Gorshelev[1], *Sergei* Orlov[1], *Oleg* Korotayev[3], *Andrei* Naumov[1,2,3], and *Aleksandr* Starukhin[4]

[1]Institute of Spectroscopy RAS, 108840 Troitsk, Moscow, Russia
[2]Moscow Institute of Physics and Technology, 141700 Dolgoprudny, Moscow region, Russia
[3]Moscow State Pedagogical University, 119991 Moscow, Russia
[4]B.I. Stepanov Institute of Physics, NAS, 220072, Minsk, Belarus

Abstract. The results of low-temperature measurements of spectral trails (temporal evolutions of individual fluorescence excitation spectra) of single molecules Mg-tetraazaporphyrin embedded in polyisobutylene are presented. Spectral trails were recorded in a broad spectral range (up to 4000 GHz) at a temperature 6K. Spectral diffusion in a broad spectral range has been found. Individual spectra of single molecules with a broad peak shifted with respect to zero-phonon line at ~14 – 15 cm^{-1} were recorded. These broad peaks can be interpreted as the phonon sidebands.

Single molecule spectromicroscopy is a quite novel but well-established experimental technique. The method is especially powerful at cryogenic temperatures when bright and narrow zero-phonon lines (ZPL) are reachable for the observation in the single molecules (SM) spectra [1,2]. Ultra-high sensitivity of the ZPLs to the SM local (nanometer-scale) environment reveals the possibility to probe solid-state media at the microscopic level [2,3]. It is well known that stochastic spectral jumps of ZPL position in spectra corresponds mainly to the coupling of impurity chromophores with tunnelling two-level systems (TLS), in turn, the electron-phonon interaction leads to the appearance of homogeneous broadening and shifts of ZPLs and gives rise to the phonon sidebands (PSB).

One of the very intriguing questions is how impurity molecules interact with so-called quasilocalized low-frequency vibrational modes (LFM). The way to study this interaction is a direct measurement of a PSB by the technique of fluorescence excitation SM spectroscopy. The PSB contains the information about LFMs energy spectra and in some cases theoretically can disclose even the individual parameters of LFMs.

Mg-tetraazaporphyrin (Mg-TAP) molecules play an important role in the photosynthesis processes and are of particular interest for studying the processes of energy storage and conversion in biological objects. Moreover, from the hole-burning experiments Mg-TAP molecules are known to be photochemically stable luminophores with narrow ZPL. In general, metal-organic molecules are very promising for nanophotonics applications [4].

* Corresponding author: savostianov.a@isan.troitsk.ru

In the present research at cryogenic temperatures (6 K) the individual spectra of single Mg-TAP molecules embedded in polyisobutylene were detected. The measurements of spectral trails in a broad spectral range up to 4000 GHz (~130 cm^{-1}) were also performed. SM fluorescence was excited by a Coherent CR-599 tunable dye laser with an effective linewidth of 10 GHz (~0.33 cm^{-1}), the scan step was 5 GHz (~0.16 cm^{-1}). The spectra were recorded in a spectral range (570 – 595 nm) corresponding to the Q-band absorption maximum of Mg-TAP molecules. The Stokes component of single-molecule luminescence was recorded in the wavelength range from 612 to 644 nm using high-sensitive EM-CCD camera (Andor iXON Ultra) [5].

The measured zero-phonon luminescence manifests different temporal dynamic and exhibits wide distribution of the values of spectral shifts in a broad range from 10 to 950 GHz (~0.33 – 31 cm^{-1}) that considerably exceeds the characteristic values observed earlier [6]. In the dipole-dipole approximation of the interaction between SM and TLS [7], the value of 950 GHz corresponds to the distance in the order of 1 nm between them. Another possible explanation of such anomalous behavior is switching between conformational substates in Mg-TAP molecules.

One of the recorded spectral trails consists not only from ZPL but also from the broad peak shifted with respect to ZPL by ~15 cm^{-1} and performing synchronous spectral jumps with ZPL. This corresponds to the peak in LFMs energy spectra – «bozon peak» that had been measured earlier in different independent experiments (see [8] and references therein). This peak can be interpreted as PSB. In another individual spectrum a quite broad peak (~4.1 cm^{-1}) shifted with respect to ZPL at 13.9 cm^{-1} was observed. This peak possibly corresponds to the interaction of a SM with a single LFM [9].

This work was done with the support of the Program of the Russian Academy of Sciences ("Actual problems of photonics"). Part of the work (A.A.S.) was done with the support of Horizon 2020 program (grant 645628).

References

1. M. Orrit, W.E. Moerner, *High Resolution Single-Molecule Spectroscopy in Condensed Matter, in Physics and Chemistry at Low Temperatures* (Pan Stanford Publishing, Singapore, 2011)
2. A. V. Naumov, Physics Uspekhi, **56**, 605 (2013)
3. K. K. Rebane, J. Lumin., **100**, 219 (2002)
4. K. R. Karimullin, A. V. Naumov, J. Lumin., **152**, 15 (2014)
5. A. V. Naumov, I. Yu. Eremchev, A. A. Gorshelev, Eur. Phys. J. D, **68**, 348 (2014)
6. Y. G. Vainer, A. V. Naumov, M. Bauer, L.Kador, Opt. Spectr., **94**, 873 (2003)
7. E. Geva, J.L. Skinner, J. Phys. Chem. B, **101**, 8920 (1997)
8. I. Y. Eremchev, A. V. Naumov, Y. G. Vainer, L. Kador, J. Chem. Phys., **130**, 184507 (2009)
9. A. V. Naumov, Yu. G. Vainer, L. Kador, Phys. Rev. B, **79**, 132201 (2009)

EPJ Web of Conferences **190**, 04020 (2018) https://doi.org/10.1051/epjconf/201819004020
HBSM-2018

Deep tumor imaging by upconversion nanoparticles

Nataly Sholina[1,2,3,*], *Polina* Demina[3], *Dmitry* Khochenkov[1,3], *Alla* Generalova[3,4], *Andrey* Nechaev[5], and *Eugeny* Khaydukov[2,3]

[1]N. N. Blokhin Russian Cancer Research Center, Moscow 115478, Russia.
[2]I. M. Sechenov First Moscow State Medical University, Moscow 119991, Russia.
[3]Federal Scientific Research Center Crystallography and Photonics RAS, Moscow 119333, Russia
[4]Shemyakin-Ovchinnikov Institute of Bioorganic Chemistry RAS, Moscow 117997, Russia
[5]Institute of Fine Chemical Technologies, Moscow Technol. University, Moscow 119571, Russia

Abstract. In this work are shown the prospects of using upconversion nanoparticles (UCNPs) as markers for contrast optical imaging of a tumor. For using nanoparticles for biomedical purposes is implemented a technique for coating nanoparticles with polymers, such as PEG and PSA. This approach provides low non-specific adsorption, which prolongs the circulation of UCNPs in mouse bearing Lewis Lung Cancer (LLC) up to 10 hours. These properties allow nanoparticles to quickly accumulate in the tumor. Effective delivery of particles with different polymer coatings in the tumor is demonstrated with the help of an epiluminescent imaging system.

Today one of the topical directions in modern medical diagnostics is the creation of nanoprobes aimed at pathological targets based on photoluminescent nanoparticles possessing high photo- and chemical stability and their application in real time. Such nanoparticles are able to specifically label tumor tissues and ensure their visualization in real time [1].

Upconversion nanoparticles (UCNPs) are a special class of luminescent nanoparticles with a specific property - the ability to conversion low-energy photons into photons of higher energy. UCNPs are inorganic nanocrystals (NaYF4) doped with lanthanide ions. Their photophysical properties provide a bright and long-term visualization of the structures marked by them in the depth of biological tissues. UCNPs are characterized by photostability, low toxicity, and their surface, allows the creation of conjugates with targeting modules. The wavelength of excitation and photoluminescence UCNPs falls into the so-called "window of biotissue transparency", which allows them to be visualized in biological tissues from a depth of several centimetre [2]. They have chemically active functional groups on the surface, which is important for obtaining targeted drugs, and their optimal size determines the predominant accumulation of nanoparticles in the tumor tissue due to the morphological features of its vascular system due to the EPR-effect. It is important to note that one of the main difficulties in the targeted delivery of nanoparticles is

* Corresponding author: sholinanv@gmail.com

their insufficient circulation time in the blood and significant absorption by macrophages, which reduces their accumulation in the tumor. To improve the targeted delivery and reduce non-specific absorption by macrophages, we modified their surface with hydrophilic biocompatible coatings based on polyethylene glycol (PEG) and polysialic acid (PSA), which increased the circulation time in the bloodstream from several minutes to 10 hours (in the case with PSA) (Fig.1A).

In order to visualize the tumor with the UCNPs with coatings, we performed a study on C57Bl/6 mice bearing Lewis lung cancer. To visualize the fluorescent signal, an imager of small animals DVS-02 (Institute of Photon Technologies of the RAS (Troitsk, Moscow)) was used. UCNPs at a concentration of 800 μg / ml in physiological saline was administered intravenously to the animals in a volume of 0.15 ml. After the administration of UCNPs, a series of epiluminescent images and spectral characteristics of the tumor were obtained to study the accumulation of nanoparticles in the tumor.

According to the obtained data, an increase in the photoluminescent signal from the tumor begins after 1 minute, with an increase in contrast as compared to the surrounding tissues. In the range of 2-5 minutes, there is a slight decrease in the signal from the tumor, which is supposed to be due to the redistribution of nanoparticles with circulating blood. Five minutes after the administration of UCNPs, the signal grows in the tumor (due to the accumulation and retention of circulating UCNPs), reaching a maximum after 60 min after administration (Fig.1B).

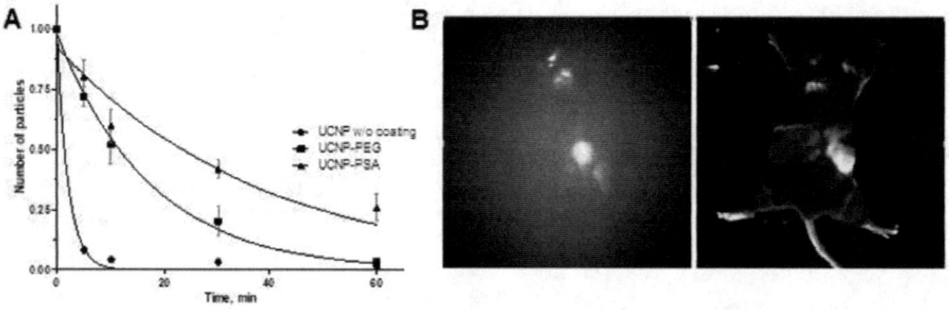

Fig. 1. (A) Normalized dependence of the UCNPs without biocompatible coating, and UCNPs surface modified with PEG and PSA vs. blood circulation time. (B) Epiluminescent images of C57Bl/6 mice bearing Lewis lung cancer obtained after 60 min UCNP-PSA injection (left) and general view (right).

The recorded photoluminescence signal was maintained for 24 hours after administration of UCNPs. The strongest photoluminescent signal was recorded from UCNPs coated with PSA, which is associated with a more significant accumulation of particles in the tumor. The smallest signal was recorded from hydrophilic UCNPs without a biocompatible coating. The conducted studies show the prospects of using nanoparticles of modified PSA for marking and intravital imaging of tumors. This technology can be in demand for intraoperative assessment of the boundaries of surgical intervention.

This work was supported by Russian Science Foundation (RSF) grant 16-13-10528.

References

1. A. Nadort, V.K.A. Sreenivasan, Z. Song, E.A. Grebenik, A.V. Nechaev, V.A. Semchishen, V.Ya. Panchenko, A.V. Zvyagin, PLoS ONE **8**, e63292 (2013)
2. B.J. Tromberg, N. Shah, R. Lanning, A. Cerussi, J. Espinoza, T. Pham, J. Butler, Neoplasia **2**, 26 (2000)

Absorption and emission properties of QD-like particles

Maxim Gladush[1,2] and *Ekaterina* Smirnova[1,3,*]

[1]Institute for Spectroscopy RAS, 108840 Troitsk, Moscow, Russia,
[2]Moscow State Pedagogical University, 119435 Moscow, Russia
[3]Moscow Institute of Physics and Technology, 142700, Dolgoprudny, Moscow region, Russia

Abstract. In this work we present a theoretical calculation of the absorption spectra of a weak probe signal and the resonance fluorescence spectra of two-level systems with a cw excitation with phonon-induced transitions.

Resonance fluorescence is the phenomenon of light emission by quantum systems in an external field. Specific spectrum known as Mollow triplet was firstly observed from Na gas [1] and later from organic molecules [2]. Today quantum dots are being studied as objects that demonstrate resonance fluorescence. In the latest experiments the resonance fluorescence including asymmetric Mollow triplets were observed [3,4].

In this study we calculated the spectra of the collective ensemble of two-level emitters. It was considered that in a dense homogeneous ensemble in a laser light each atom is driven by an effective field. The work [5] was taken into account to consider the collective relaxation rate. A characteristic feature of the fluorescence spectrum of this system is that the side peaks in the Mollow triplet are spectrally moved from the center.

To simulate a resonance fluorescence spectrum of a single QD the model of quantum emitter was considered with the mechanisms of quantum transitions between its states and processes of cross-relaxation of coherences due to the electron-phonon interaction. The possibility of asymmetric Mollow triplet was demonstrated. The assumptions about the model of quantum dot were taken from work [4].

We carried out simulations of resonance fluorescence spectra and the probe signal absorption spectra of an ensemble of QD-like particles and demonstrated the possibility of the appearance of an asymmetric Mollow triplet with positions of the side peaks strongly dependent on laser detuning from the frequency of the transition in the emitter. This result qualitatively corresponds to the spectral dependences recorded in experiments with single quantum dots.

This research was supported by Russian Science Foundation (№ 17-72-20266).

References

1. F. Y. Wu, R. E. Grove, S. Ezekiel, Phys. Rev. Lett., **35**, 1426 (1975)

[*] Corresponding author: ksmirno@yandex.ru

2. G. Wrigge et al., Nature Physics, **4**, 60 (2008)
3. Y. J. Wei et al., Phys. Rev. Lett., **113**, 097401 (2014)
4. A. Ulhaq et al., Opt. Express, **21**, 4382 (2013)
5. M. G. Gladush, D. V. Kuznetsov, V. K. Roerich, Eur. Phys. J. D, **64**, 511 (2011)

EPJ Web of Conferences **190**, 04022 (2018) https://doi.org/10.1051/epjconf/201819004022
HBSM-2018

Nanohybrid scaffolds with luminescent remote control

Anastasia V. Sochilina[1,2,*], *Alexander* G. Savelyev[2,3], *Natalya* V. Sholina[3,4], *Denis* N. Karimov[2], *Andrey* V. Nechaev[2,5], *Evgeny* V. Khaydukov[2,3], and *Alla* N. Generalova[1,2]

[1]Institute of Bioorganic Chemistry RAS, 117997, Moscow, Russia
[2]Scientific Research Centre "Crystallography and Photonics" RAS, 119333, Moscow, Russia
[3]Institute of Molecular Medicine, Sechenov First Moscow State Medical University, 119991 Moscow, Russia
[4]N.N. Blokhin Russian Cancer Research Center of Ministry of Health of the Russian Federation, 115478, Moscow, Russia
[5]Institute of Fine Chemical Technologies, Moscow Technological University, 119571, Moscow Russia

Abstract. We report on hybrid nanocomposite scaffolds on the base of cross-linked hyaluronic acid derivative with embedded upconversion nanoparticles (UCNPs). The unique photoluminescence properties of specially designed hydrophilic UCNPs enable visualization of hydrogel using NIR irradiation. Formation of scaffold structure can be produced by means of 3D printing or direct laser writing. For the first time, we present visualization of nanohybrid scaffolds in live small animal aiming to demonstrate new possibilities of their luminescent remote control for tissue engineering.

1 Introduction

Hyaluronic acid (HA) appears to be one of the most suitable materials for tissue engineering and regenerative medicine. This naturally occurring polysaccharide is *a priori* biocompatible and biodegradable, still it lacks enough mechanical strength for constructing scaffolds. One of the methods to reinforce this polymer is photocrosslinking of HA derivatives containing double bonds. Addition of photoinitiator into hyaluronic acid glycidyl methacrylate (HAGM) water solution and irradiation it with UV/blue light makes possible crosslinking of composition resulting in solid insoluble hydrogel formation. Incorporation of UCNPs composed of $NaYF_4$ crystals codoped with Yb^{3+} and Er^{3+} enables visualization of produced hydrogels under NIR light excitation as these UCNPs exhibit anti-Stokes luminescence under excitation at 980 nm. Since NIR light falls in so called "transparency window" for biological tissues, it can penetrate tissue and organs up to centimetres unlike visible light.

2 Results and discussion

In this work double bond moieties were introduced into HA by conjugation of the polysaccharide with glycidyl methacrylate (GMA) producing derivative capable of crosslinking (HAGM). Conjugation of GMA can be carried out via reaction mechanisms of

* Corresponding author: ddraig@yandex.ru

reversible transesterification or irreversible epoxy ring opening with carboxyl or hydroxyl moieties of HA [1]. In order to make core/shell $NaYF_4:Yb^{3+}/Er^{3+}$ UCNPs soluble in water their surface was first covered with amphiphilic SPAN-60 (sorbitan monostearate), then with polyethyleneimine for biocompatibility. HAGM (20 wt.%) was further dissolved in water solution containing UCNPs with following addition of polyethylene glycol diacrylate (PEGDA), Mn=575 (5 wt.%), and flavin mononucleotide (0,0004 wt.%) as an endogenous photoinitiator [2].

The produced photocurable composition (PCC) was subjected to exposure at 365 nm wavelength to gelate 150 µm thin hydrogel films 6 mm in diameter. The acquired scaffolds were implanted subcutaneously on dorsal side of mice with scaffolds without UCNPs as control. Animals were monitored with bioimaging system every day (Fig. 1, left). Intravital luminescent signal under NIR light irradiation was recorded up to 7th day after, then strong signal was recorded *post mortem* from extracted scaffold. Histological analysis displayed weak inflammation reaction and slight degradation. As an example of structured scaffold we produced the structure using PEGDA-based PCC by means of direct laser writing (Fig. 1, right).

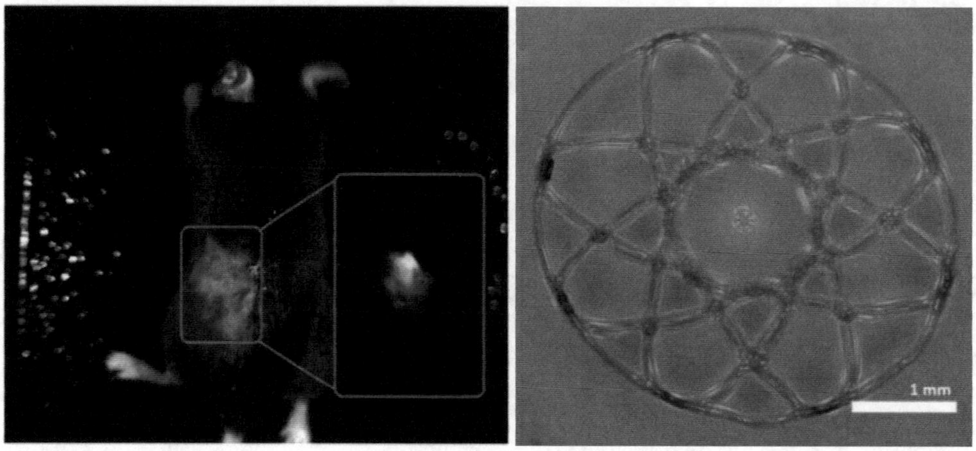

Fig. 1. Left: Bright-field photo of mouse with implanted nanohybrid scaffold on day 1. The inset shows epiluminiscent signal from scaffold under irradiation at 980 nm wavelength. Right: Scaffold structure produced by direct laser writing.

This scientific work has been partly supported by the Federal Agency of Scientific Organizations (Agreement No 007-ГЗ/Ч3363/26) in the part of scaffold formation and Grant RSF No. 18-79-10198 in the part of modification of UCNP surface.

References

1. S. A. Bencherif, A. Srinivasan, F. Horkay, J. O. Hollinger, K. Matyjaszewski, N. R. Washburn, Biomaterials, **29**, 1739 (2008)
2. A. G. Savelyev, A. V. Sochilina, R. A. Akasov, A. V. Mironov, V. A. Semchishen, A. N. Generalova, E. V. Khaydukov, V. K. Popov, Sovremennye tehnologii v medicine, **10**, 88 (2018)

Spectral features of conjugated bichromophores: porphyrins with curcumin

Aleksandr Starukhin[1,*], *Aleksander* Gorski[2], *Tatsiana* Pavich[1], and *Valerii* Kniukshto[1]

[1]B.I. Stepanov Institute of Physics, National Academy of Sciences of Belarus, Minsk 220072, Belarus
[2]Institute of Physical Chemistry, Polish Academy of Sciences, Warsaw 01-224, Poland

Abstract. Conjugated bichromophores of porphyrins with natural compound curcumin were synthesized and studied by methods of spectroscopy. The fluorescence spectra of bichromophore consist of four bands covered almost a whole visible spectrum from 450 to 700 nm. A relatively weak interaction between curcumin and metalloporphyrin allows to combine both effects: high generation efficiency of singlet oxygen by metalloporphyrin and an anticancer treatment of curcumin.

Design and characterization of novel types of polyheterocyclic compounds have a huge impact on development of methods for the selective transportation of photosensitizers to tumour cells,- a crucial point of photodynamic therapy. Well known, that porphyrins are widely used as effective photosensitizers of singlet oxygen generation in photodynamic processes [1,2]. The idea to design a conjugated bichromophore based on porphyrin macrocycle with a natural compound - curcumin, isolated from the Curcuma longa plant, seems very promising. Nowadays curcumin isomers and derivatives attract a lot of attention of specialists in the field of medicine and biochemistry, primarily as an effective antitumor drugs [3]. Number of publications devoted to the study of the optical and photophysical properties of curcumin [4], as well as its structure modifications have been appeared resently.

In this report we presented of the first results of the synthesis of a structure based on metallocomplexes derivatives of tetraphenylporphyrin linked to curcumin molecule and studying of optical parameters newly-synthetized compounds at ambient and liquid nitrogen temperatures. The structures of Zn-5-(4-aminophenyl) 10,15,20-(triphenyl) porphinato (ZnATPP) and Zn-5-(4-aminophenyl)-1-(4-N-curcumin) 10,15,20-(triphenyl) porphinato (ZnATPP-CURC) are presented on Fig. 1 (a, b).

Fluorescence excitation, fluorescence and phosphorescence spectra of ZnATPP and ZnATPP-CURC have been recorded at ambient and liquid nitrogen temperatures. Fluorescence excitation and fluorescence are presented on Fig. 2 (a, b). As evidently from absorption and fluorescence excitation spectra (Fig. 2 a) absorption bands of curcumin and ZnAPTPP are practically coincided in the spectral range of 400-500 nm.

* Corresponding author: starukhin@mail.ru

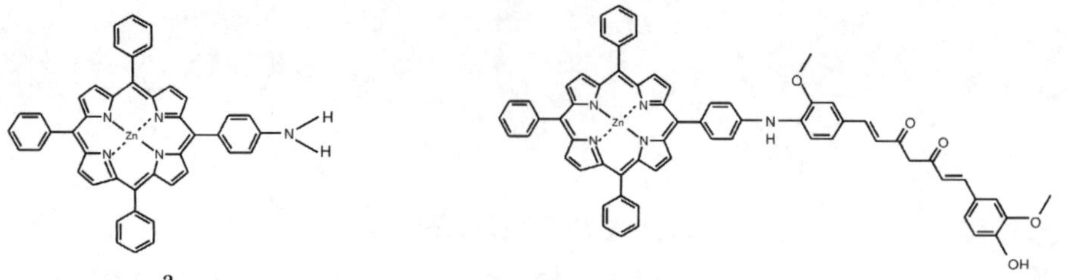

a b

Fig. 1. Chemical structures of ZnATPP (a) and ZnATPP-CURC (b) compounds.

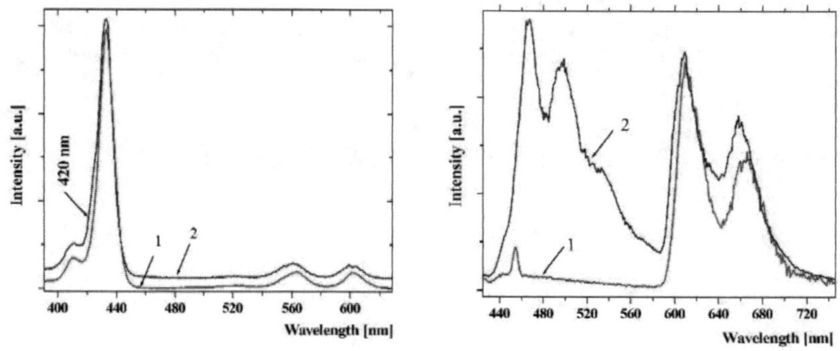

a b

Fig. 2. Fluorescence excitation (a) and fluorescence (b) spectra of ZnATPP (1) and ZnATPP-CURC (2) in toluene at 77 K.

The absorption bands of ZnATPP-CURC bichromophore exhibits a minor difference compare to ZnATPP in the range of 420 nm. While, the fluorescence spectrum of ZnAPTPP consists from two bands similar to well-known ZnTPP. In contrast, four bands in the spectral range from 450 to 700 nm (Fig. 2b.2) have been observed in the fluorescence spectrum of ZnAPTPP-CUCR. Two "blue" bands in the spectrum (at 465 and 497 nm) refer to the curcumin emission, and two "red" bands at 608 and 659 nm correspond to the luminescence of ZnAPTPP. Upon formation of bichromophore with curcumin, the bands in the fluorescence spectrum of ZnAPTPP are shifted by more than 2 nm to the range of short wavelengths. A relatively weak phosphorescence of ZnAPTPP and ZnAPTPP-CUCR were detected with a maximum at about 790 nm.

The complex of ZnAPTPP-CUCR combines a high efficiency of singlet oxygen generation by one part of the complex (ZnAPTPP) with an anticancer effect of curcumin, that can provide to high efficiency of these species as therapeutic agents in the treatment of oncological diseases.

This work was supported by European Union's Horizon 2020 research and innovation program, grant agreement № 645628 and the Belarusian-Polish project № F18PLSHG-003.

References

1. A. Krasnovsky jr., Biofizika USSR **21**, 748 (1976)
2. F. Wilkinson, W. Phillips, J. Phys. Chem. Ref. Data Rev. **22**, 113 (1993)
3. V. B. Patel et al., Nutrition and Cancer **62**, 958 (2010)
4. K. I. Priyadarsini, J. Photochem. Photobiol. C: Photochem. Rev. **10**, 81 (2009)

 https://doi.org/10.1051/epjconf/201819004024

Room temperature optical thermometry based on the luminescence of the SiV defects in diamond

Caius Miller[1], *Laurits* Puust[1], *Valter* Kiisk[1], *Evgeny* Ekimov[2], *Igor* Vlasov[3], *Yurii* Orlovskii[1,3], *and Ilmo* Sildos[1,*]

[1]Institute of Physics, University of Tartu, W. Ostwaldi Str. 1, EE50411 Tartu, Estonia
[2]Institute for High Pressure Physics RAS, Kaluzhskoe Road 14, Troitsk, Moscow 108840, Russia
[3]Prokhorov General Physics Institute RAS, Vavilov Street 38, Moscow 119991, Russia

Abstract. SiV-containing microcrystals of diamond are synthesised by using high-pressure high-temperature treatment of a mixture of pertinent organic-inorganic precursors. Photoluminescence of SiV defects were investigated with the aim to use the microcrystals for optical temperature sensing in near infrared at room temperature based on temperature-dependent shift of the 740 nm zero-phonon line of SiV photoemission.

1 Introduction

Biomedical applications require nontoxic photoresistant optical markers absorbing and emitting in the NIR biological window of biotissues. Several impurities in diamond, such as SiV and NE8 defect centres with spectrally narrow zero-phonon lines (even at room temperature), satisfy these demands [1,2].

2 Experimental

Diamond microcrystals (typical size ~1 μm) containing SiV defects were synthesized from a mixture of adamantane ($C_{10}H_{16}$) and tetraphenylsilane ($C_{24}H_{20}Si$) in a ratio of Si/C 0.28%. An apparatus of uniaxial compression and the high-pressure chamber of "Toroid-15" type were used for the synthesis [3]. The diamond sample was synthesized in titanium capsule at a pressure of about 9 GPa and a temperature of 1600 °C, the exposure time at the synthesis temperature was 10 seconds.

Under 488 nm laser excitation the SiV defects possessed an intense emission spectrum with a quite sharp zero-phonon line (ZPL) at 740 nm over a wide temperature range (Fig. 1). The spectral position of the ZPL shows a strong red-shift by increase of temperature (Fig. 2). Acceptable fitting could not be achieved by the conventional T^4 dependence but required additional T^2 term to take into account softening of bonds in the excited electronic state in accordance of a modified model of the ZPL of impurity centers in solids [4].

* Corresponding author: ilmo.sildos@ut.ee

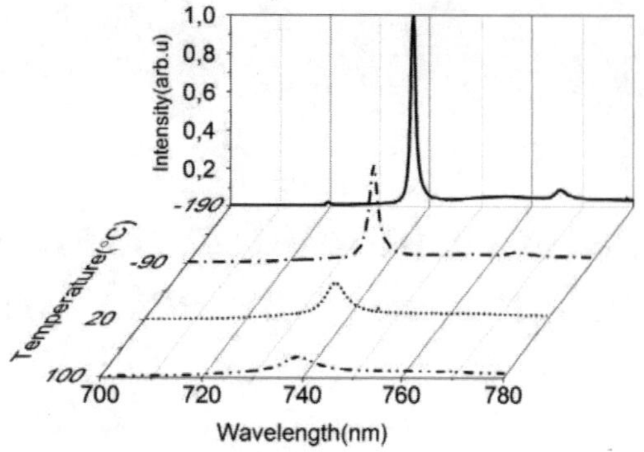

Fig. 1. Temperature dependence of SiV emission in a diamond microcrystal. Excitation with 488 nm Ar^+ laser line

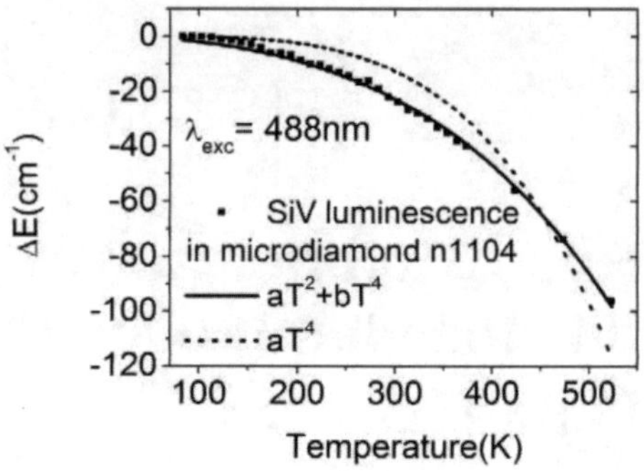

Fig. 2. Thermally induced shift of the 740 nm ZPL in the emission spectrum of SiV defects in diamond micro-crystals.

3 Concluding remarks

A narrow 740 nm ZPL and its strong position dependence on temperature are promising for a sensitive (~1 K) optical thermometry around RT. Preparation of SiV-containing nano-diamonds by using the above described HPHT technique is in progress. We further intend to test the thermo-sensing capabilities of the nanoparticles in aqueous media and implement imaging and thermometry with different cell cultures.

The research of Estonian and Russian authors was supported by EU through the ERA.Net Rusplus project "88 DIABASE" and by Russian Science Foundation (grant No 14-12-01329), respectively.

References

1. Ch. T. Nguyen, R. E. Evans, A. Sipahigil, M. K. Bhaskar, D. D. Sukachev, V. N. Agafonov, V. A. Davydov, L. F. Kulikova, F. Jelezko, M. D. Lukin, arXiv:1708.05419v1 (2017)
2. I. Sildos, A. Loot, V. Kiisk, L. Puust, V. Hizhnyakov, A. Yelisseyev, A. Osvet, I. Vlasov, Diamond and Related Materials, **76**, 27 (2017)
3. E. A. Ekimov, O. S. Kudryavtsev, N. E. Mordvinova, O. I. Lebedev, and I. I. Vlasov ChemNanoMat, **4**, 269 (2018)
4. V. Hizhnyakov, J. Phys. Cond. Matt., **24**, 104011 (2012)

EPJ Web of Conferences **190**, 04025 (2018) https://doi.org/10.1051/epjconf/201819004025
HBSM-2018

Producing single-photon wave packets of the infrared spectrum range with an orbital angular momentum using "vortex" phase plates

D.A. Turaykhanov[1], A.V. Shkalikov[1,*], A.A. Kalachev[1], I.R. Imangulova[2], N.N. Losevsky[2], E.V. Razova[2], S.A. Samagin[2], S.P. Kotova[2]

[1]Zavoisky Physical-Technical Institute, Kazan Scientific Center RAS, 420029 Kazan, Russia
[2]P.N. Lebedev Physical Institute RAS, Samara Branch, 443011 Samara, Russia

Abstract. We consider the peculiarities of formation and registration of axially symmetric vortex fields in the context of applying diffractive optical elements such as vortex lenses and vortex axicons for the generation of single-photon states with a nonzero orbital angular momentum in the process of spontaneous parametric down-conversion.

For the realization of free-space quantum communication channels, light beams with an orbital angular momentum (OAM) are of greatest interest [1, 2]. Experimental and theoretical studies show that such beams are more stable when propagating in a turbulent medium compared to ordinary Gaussian beams [3, 4], and exhibit self-healing ability [5, 6].

Single-photon states at a wavelength around 800 nm corresponding to one of the atmospheric transparency windows can be effectively obtained in the process of spontaneous parametric down-conversion (SPDC) of light in a nonlinear medium [7]. In this paper, we present the results of preliminary studies devoted to generation of single-photon states with a nonzero OAM in the process of SPDC with the use of diffractive optical elements.

To create optical vortices we used diffractive optical elements (DOE) in the form of a vortex lens and a vortex axicon [8, 9], which is one of the most flexible and energetically effective methods. We studied theoretically and experimentally the evolution of the generated fields during their propagation near the focusing region for a vortex lens and a vortex axicon. Fig. 1 shows the diameter of the beam ring as a function of the distance along the propagation axis. It can be seen that for the axicon the distance along the propagation axis, over which the intensity distribution remains practically constant, is about 2 times larger than that for the vortex lens. For the mask parameters used in the experiment, it is about 4 cm and 2 cm, respectively. On the other hand, the vortex lens provides a greater contrast of intensity distribution in the focal plane, which can be important when detecting the output signal. Therefore, when creating an experimental setup, the problem of choosing between vortex lenses and vortex axicons should be solved taking into account the geometry of the pump field and the geometry of the registration scheme.

* Corresponding author: andrei_vs@rambler.ru

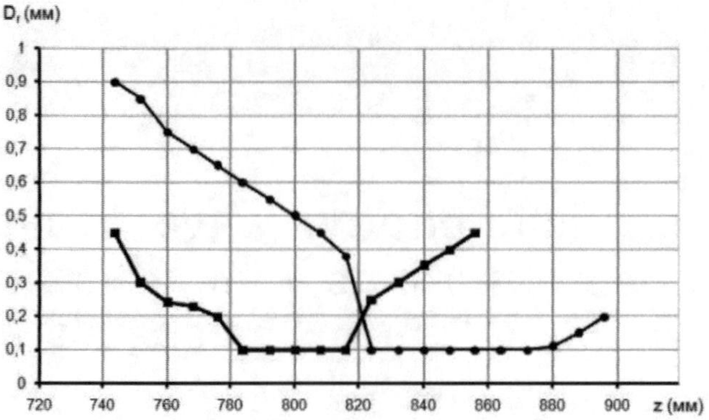

Fig. 1. The width of the beam ring as a function of the distance in the focus region for the axicon (circles) and for the vortex lens (squares).

One of the most effective methods for generating single photons is the process of SPDC, during which a photon from a strong pump field propagating in a non-centrosymmetric medium is annihilated with creating two photons, called idler and signal. This process satisfies the energy and phase matching conditions: $\boldsymbol{\omega_p} = \boldsymbol{\omega_i} + \boldsymbol{\omega_s}$ and $\vec{k}_p = \vec{k}_i + \vec{k}_s$, respectively, where ω and $\vec{k}$ are the frequency and wave vector, respectively, of the signal (i), idler (s) and pump (p) photons. In addition, for the collinear SPDC the following condition is valid for OAM: $\ell_p = \ell_s + \ell_i$. We numerically simulated the collinear SPDC process for $\ell_p = \mathbf{-1, 0, +1, +2}$, and calculated coincidences between the signal and idler photons for 49 possible combinations of orthogonal states. As should be expected, the coincidences observed in all cases when the sum of OAM of the generated photons is equal to that of the pump beam.

The paper presents the results of experimental studies devoted to the generation of single photons with a non-zero OAM via SPDC. We consider the peculiarities of formation and registration of axially symmetric vortex fields with the use of diffractive optical elements: vortex lenses and vortex axioms. It was found that the position of the detection plane in the case of vortex lenses is more critical than in the case of vortex axioms. Compensatory method of measuring the topological charge allows one to achieve at least a tenfold increase in the signal compared to the method without compensation.

This work was supported by the Russian Foundation for Basic Research (project No. 16-02-01055-a) and by the theme of the state assignment No. 0217-2018-0005 (numerical simulations of the SPDC process).

References

1. A.M. Yao, M.J. Padgett, Adv. Opt. Photon., **3**, 161 (2011)
2. J. Wang, Photon. Res., **4**, B14 (2016)
3. C. Paterson, Phys. Rev. Lett, **94**, 153901 (2005)
4. A. Vinotte and L. Berg, Phys. Rev. Lett, **95**,193901 (2005)
5. Z. Bouchal, Opt. Commun, **210,** 155 (2002)
6. M. V. Vasnetsov et al., JETP Lett, **71,** 130 (2000)
7. D.N. Klyshko, Photons and Nonlinear Optics, Gordon & Breach (1988)
8. E. Abramochkin et al., Optics Comm, **102,** 336 (1993)
9. A.V. Shkalikov et al., Bull. Lebedev Phys. Inst., **45,** 79 (2018)

AUTHOR INDEX

AUTHOR INDEX